THE WAR GAME

War Games photographed by Philip O Stearns

The WAR GAME

General Editor
Peter Young
DSO MC MA FSA FR HistS FRGS

CASSELL LONDON

CASSELL & COMPANY LTD

35 Red Lion Square London WC1R 4SJ
Sydney, Auckland, Toronto, Johannesburg

First published 1972

Made by Roxby Press Productions
55 Conduit Street London W1R 9FD

Editor Michael Leitch
Picture research Penny Brown
Design and Art direction Ivan and Robin Dodd
Model soldier consultants Hinchliffe Models
All terrain made and supplied by Peter Gilder
of Hinchliffe Models

Printed in Great Britain by Oxley Press Limited

F.672/ISBN 0 304 29074 2

Contents

PREFACE

by Aram Bakshian, Jr, Editor of The Vedette,
Journal of the National Capital Military Collectors
Past Commanding Officer, National Capital Military Collectors

It is a sad commentary on human nature that throughout history the sword has been the final arbiter. Yet out of the horror and carnage of war have come some of the most heroic chapters of the human story, some of its greatest leaders and greatest lessons. Assembled in these pages are highlights of more than two millenia of military history: battles that changed history, for better or for worse. The editor, Brigadier Peter Young, is a man equally deft with the pen and the sword. He has known both the grim business and the colourful game of war, and he has assembled a team of knowledgeable, articulate writers who have brought alive its splendour, its squalor and its undeniable fascination.

Here is vibrant, living history that reconstructs and animates, that tells us *how* and *why* momentous results were achieved on the field of battle, whether at ancient Thermopylae or El Alamein of living memory.

For thousands of war-gamers, this book provides an invaluable guide to re-enacting great moments of military history on the bloodless battlefield of the tabletop. For the military collector, it offers lavish illustrations of some of the finest miniatures, paintings, models, maps and dioramas ever assembled under one cover. And, for the general reader, here is an engrossing glimpse of mankind's most dazzling, most dangerous game: the war-game, as it has been fought and played over the ages.

March 1972 A. B., Jr
Washington D.C.

INTRODUCTION

Although wars continue to be waged throughout the world, the invention of nuclear weapons has made them too destructive and too expensive to be a useful means of furthering state policy. But the more deadly and the more costly war becomes, the more the reading public becomes interested in military history. The lives of great commanders, the uniforms of famous regiments, the histories of old corps, which before the Second World War attracted only specialists and veterans, now seem to have an almost universal appeal. It is natural that man should take an interest in something that may carry him off in his prime. Its study is a kind of ju-ju. It is as if a man feels that by studying war he can avert it.

This book is concerned with the strategy and tactics of land warfare through the ages. It describes in close detail ten important battles and can be enjoyed both as an impressive and fascinating study of the art of war and as a guide for war-gaming enthusiasts wishing to work out their own solutions on their own war-game tables. Tactics depend to a great extent on numbers, on weapons and upon terrain. For this reason the authors have paid great attention to the assets of each side – to the order of battle – and to the maps.

The book has a fresh approach. It serves two aims: firstly, to inform and entertain anyone who has an interest, however remote, in the motivation, the politics, the pageantry, the skill and courage of our ancestors, who at decisive moments in the past, hammered history into shape. Secondly, these ten accounts will, it is hoped, prove a valuable reference source for historians and war-gamers alike. The Prussian army pioneered the *Kriegspiel*, or war game, as early as 1824, and derived great advantage from it. Since those days the British and American armies have also made use of it for both planning and instruction.

The aims of this book are neither so ambitious nor so serious. Nevertheless, it may foster an interest not only in the fascinating – and relatively bloodless – hobby of war-gaming, but also in the history of the art of war. It is hoped that the student will derive special benefit from the documentary evidence on which the book is based, drawn as it is from the annals of old wars. At the same time, contemporary evidence is seldom exhaustive. Not every general wrote his memoirs; not every adjutant kept up the regimental war diary; the despatches of defeated commanders are not always in evidence. Who has ever seen the Saxon eye-witness account of Hastings? Some at least of the problems of military history can be solved, as the late Lieutenant-Colonel Alfred Burne liked to, by Inherent Military Probability (IMP). Here the war-game with its insistence on representing all the units and formations present makes a most useful guide. How did Sir William Balfour make his charge at Edgehill? If you have laid out your battlefield properly the answer will hit you in the eye – or at least it should do. The old-fashioned, dry-as-dust historian who seldom stirred from his study, far less stumped in all weathers across the stricken fields of times gone by, was ill-equipped to work things out in the detail beloved, and rightly so, of the enthusiast determined to reach the heart of the matter.

The Choice of Battles

The book ranges widely over the field of military history, and this is justifiable at a time when the ancients, with their bows, slings and spears, are becoming ever more popular with war-gamers. Even so, we have concentrated on the musket period from Marlborough to Wellington, when the teeth of any army consisted of horse, foot and guns, and tactical problems had a beautiful simplicity. Close-order drill and short-range weapons are ideal for tactical war-games, if only because the player must be able to reach the middle of the table in order to manoeuvre his soldiery. For this reason we have included only one modern battle. The dispersal required by modern weapons demands a war-game room the size of a drill hall! From the visual point of view, too, there is little to recommend modern warfare. Perhaps it would have been prudent had the politicians abolished war when the soldiers introduced khaki!

It would be invidious to extol here the merits of the contributors who have collaborated to bring these battles to life. Suffice it to say that they were invited to join the team not only for their experience as writers on military topics, but because they are all veterans of war-games too numerous to name. In addition, not a few of them have hazarded their persons on fields every bit as dangerous as a table top – in France, Burma, Africa and elsewhere.

Ripple

Peter Young

Charles Grant at

THERMOPYLAE

Betrayed by the traitor Ephialtes, who guided a force of Persian Immortals' through the mountains and behind the Greek position, deserted also by their fellow Greeks in their hour of need, Leonidas and his 300 Spartans drew close their shields for the last time and faced annihilation. To a man they died in the Pass of Thermopylae, which they and others had so bravely defended for two days against the frontal assault of mighty Xerxes and his Persian army.

480BC

Part of the Greek advance party led by Leonidas arrives in the Pass of Thermopylae, a near-impregnable position on the Malian Gulf. There the Greeks planned to hold the Persian army while the Greek fleet met and attempted to destroy the Persian navy, whose vessels were flanking and supplying the land forces.

The Background to the Battle

In the fifth century BC Persia had extended her borders as far as they would go to the north and east. She then perceived that the islands and mainland of Greece might well provide room for expansion, and first one, then another Persian king made the attempt, approaching the Greek mainland by way of the Hellespont with the aim of pushing into Thrace and Thessaly. It was Darius I (*c.* 550–*c.* 485 BC) who made the first determined effort. Taking advantage of what he thought to be a period of inter-city discord in Greece, he launched a large-scale amphibious operation that culminated in the landing of a powerful army at Marathon. There his troops were attacked and completely defeated by a much smaller Athenian army under Miltiades. This defeat was a profoundly serious reversal for Persia and had far-reaching repercussions. Domestic revolts broke out in the Empire and these ensured that a large part of the Persian army was kept busy for some years within the Imperial borders. Indeed, after Darius's death his son Xerxes (*c.* 519–*c.* 465 BC) was also much occupied with putting down risings in various parts of Persia.

To the Greeks, however, it was obvious that this respite was to be a short-lived one, and in fact by 484 BC Persian preparations for a massive invasion of Greece were well in hand. It was the Athenian statesman Themistocles who seemed most acutely aware of the thunderclouds looming over Asia Minor, and it was largely through his personal endeavours that active efforts were made to bring the Athenian fleet – a most vital factor in the defence of Greece – up to fighting strength.

Meanwhile Persian preparations proceeded at a furious rate. Two bridges of boats were flung across the one-mile-wide waters of the Hellespont; ships were gathered from every port of Asia Minor and Egypt, and vast if somewhat heterogeneous forces were assembled from all the huge areas owing allegiance to Xerxes, the 'Great King'. Finally, early in the year 480 BC, Xerxes led his army from Sardis in Lydia northwards to the crossing-point, and by the middle of May an enormous array of men, infantry, cavalry and baggage trains had been transferred into Europe and was streaming across Thrace towards Thessaly and Greece. Contemporary accounts give hugely inflated numbers for the Persian army, but it probably numbered some 150,000 men (although this is a conjectural figure).

Despite a lasting inability to agree among themselves the Greek states, or at least those which had no intention of acknowledging the sovereignty of Persia, decided upon a common defence policy. The cornerstone of this policy was to be a holding action at the Pass of Thermopylae, through which Xerxes was expected to bring his army. This would allow the Greek fleet to meet and, it was hoped, destroy the naval might of the Persian king, whose vessels were flanking and supplying the land forces in their progress.

The force destined for this unpromising duty – an almost hopeless cause in view of the numbers of the enemy – was

MAP 1 THE THEATRE OF WAR

This map shows the invasion route from Sardis of Xerxes' massive Persian army, and also the line of march followed by Leonidas, the Spartan king, with a numerically much inferior force. The Greek army chosen to defend Thermopylae was initially drawn from the Peloponnese and consisted of some 4,000 men (of whom only 300 were Spartans). In the course of his march northwards, Leonidas secured additional troops from various neighbouring states until his army was about 8,000 strong – though that was still a pitifully small number with which to resist some 150,000 Persians.

drawn from the Peloponnese, and ideally should have been largely Spartan. But of the 4,000 men who marched northwards to take up the position only 300 were citizens of that most martial of Greek cities. The Spartans had claimed that an important religious festival prevented them from mustering troops, but they had promised that the full might of the city would follow when the festival was concluded. (It must be said that, while the Spartans were certainly a devout people, this was not the first time that religious functions had been given precedence over military operations.)

In any event the advance guard – if it can be so described – was commanded by one of the two kings of Sparta, Leonidas. Having passed the Isthmus of Corinth on his march north, he received some most welcome additions to his army from various cities which had also decided not to throw in their lot with Persia. Seven hundred men came from Thespiae, there were 400 Thebans, and 1,000 each from Phocis, Malis and Locris. So, on Leonidas's arrival at the Pass of Thermopylae, his united army numbered some 8,000 men – although not all were of the same fighting quality as the Spartans.

The pass through which the Persians hoped to debouch into Greece appeared to offer a near-impregnable position. On one side cliffs rose sheer and forbidding; these afforded passage to neither man nor beast and merged with the towering mountain massif to the south. Stretching into the distance on the other side lay the waters of the Malian Gulf. The level ground seems to have been no more than twenty to thirty yards in width at the narrowest point, and there it was that the hosts of Persia were to be confronted by the bronze-panoplied men of Greece.

However strong a position may be, it is nevertheless a rare one whose flanks are absolutely secure; and at Thermo-

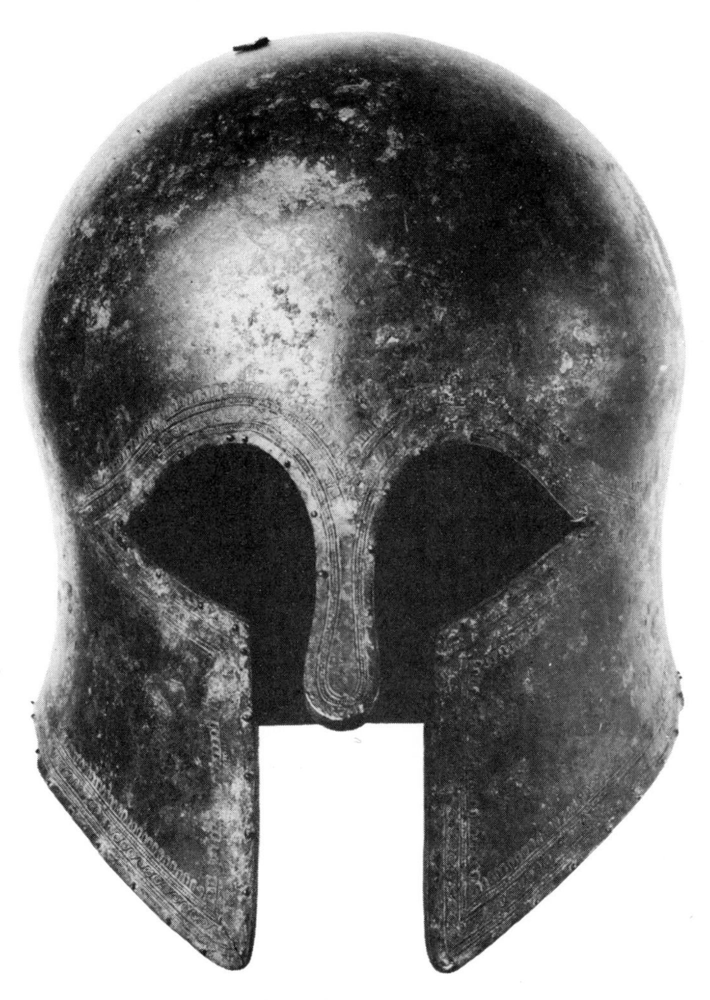

above An enclosed Greek helmet from the beginning of the fifth century BC. **below** Greek fighting men of the period; at that time the most successful fighting unit was the hoplite, a heavy infantryman armed with a round shield about three feet in diameter, a bronze helmet, a cuirass (body armour) and greaves (leg guards). He usually carried a long spear and a short sword.

pylae, despite the apparently impassable peaks and mountains to the south, there was in fact a path winding through the fastnesses, difficult indeed, but quite practicable for an active and determined infantry. Leonidas was soon told of its existence and he took urgent measures to ensure that it would be covered, posting a force of about 1,000 men to guard it (the contingent from Phocis). While these and other defensive measures were being carried out the Persian tide flowed inexorably onwards, and on 14 August its leading elements arrived at the pass. There, as their scouts had forewarned them, Xerxes' men found the road to Athens barred by the shields and spears of Sparta and her allies.

The Rival Armies

The comparative strengths of the two armies at Thermopylae were approximately as follows:

Greeks	
Spartans	300
Other Peloponnesian troops	3,700
Thespians	700
Thebans	400
Phocians	1,000
Malians	1,000
Locrians	1,000
Total	8,100

Persians[1]	
Immortals	10,000
Other Persian troops, Medes, Dahae, Scythians, etc.	140,000
Total	150,000

above A line of Persian infantrymen armed with the large but lightweight wood or wicker shields that, at Thermopylae, gave inadequate protection against the heavy thrusting spears of the Greek hoplites. **right** An archer in a frieze showing the Persian Royal Guard, from the Palace of Darius at Susa, *c.* 500 BC. The bow was a favoured weapon among the Persians, who developed a powerful composite type made of laminated wood, glue and horn.

From Persia came a mixture of every kind of fighting man that could be called to arms. In quality they ranged downwards from the fighting troops of the Great King's élite force, the 'Immortals', so named because any man lost in battle was immediately replaced; thus in theory the unit never fell below its establishment of 10,000 men. After the Immortals the Persian army spanned the entire military spectrum, ending with semi-savage and no doubt unwilling levies taken from the most distant territories of the Empire; they included Indians, nomadic Asian tribesmen and painted Ethiopians.

Possibly the best part of the Persian army was the cavalry. This consisted of troops led by lordly barons and wild tribesmen bred to the saddle in the great northern plains, among whom were the Dahae and the Scythians. Of course, in the situation at Thermopylae with mountain gorges and narrow passes to negotiate, the cavalry had no immediate value.

[1] It should be pointed out that Persian strengths at Thermopylae are computed from exclusively Greek sources, in whose interest it was to magnify the size of the enemy.

Nevertheless, as an infantry force the Immortals were fighting troops of a very high calibre. As befitted such an esteemed unit, they were equipped far more lavishly than their fellows with elaborate robes, silver-bladed spears and other richly worked items. All the contemporary representations of this famous unit, incidentally, show them in what might be termed their parade dress, but it seems probable that in action they wore coats of mail (probably of bronze) under their dashing robes. Soldiers in other infantry units had no such armour, however, and generally speaking the Persian troops had little protection. Their shields, even those of the Immortals, were made of light wood or wicker, and though good for mobility and ease of handling were incapable of withstanding violent treatment. The characteristic weapon of the Persian army was the bow. The Persians used a powerful composite type, made of laminated wood, glue and horn and capable of propelling an arrow with great force for a tremendous distance. Apart from the bow the most favoured weapon was a short spear, light in weight and not by any means in the same class as the long and heavy thrusting spear of the Greeks.

By the fifth century BC the typical Greek soldier was the hoplite, a heavy infantryman armed with a great round shield some three feet in diameter, known as the 'hoplon' (whence the soldier's name was derived). He also wore an enclosed bronze helmet, together with a cuirass (body armour) and greaves (leg guards) for added protection. The hoplite's weapon was a long spear, eight to nine feet in length, and a short sword was also carried. The city states depended almost exclusively on this type of soldiery, except for those in the north which favoured a high proportion of cavalry for use on the local plains. In most cases warriors were men of citizen status who provided their own arms and armour.

In battle the hoplites moved in solid, well-disciplined bodies called phalanxes, drawn up in lines of up to eight men in depth. The phalanx was indeed a formidable force. In the terrain where the Greeks customarily fought – narrow valleys lined by rocky slopes and steep cliffs – the hoplite phalanx, advancing with its men in close order, each man's shield overlapping that of the man on his right, its front bristling with a multitude of spear points, caused great apprehension among enemy forces unaccustomed to this mode of warfare.

The weakness of the system was that in open country, above all when fighting a mobile and nimble foe, the flanks and rear of the phalanx lay open to attack. This led in time to the development of light troops to support the phalanx; these included javelin men and archers, who could be moved about quickly to counter flanking moves.

Such developments lay largely in the future, however, and it was the archetypal hoplite who guarded the narrow passage at Thermopylae. There the Greeks under Leonidas were a mere fraction of the Persian host in number, but in the main they were well disciplined, well armed and fiercely determined to do their duty.

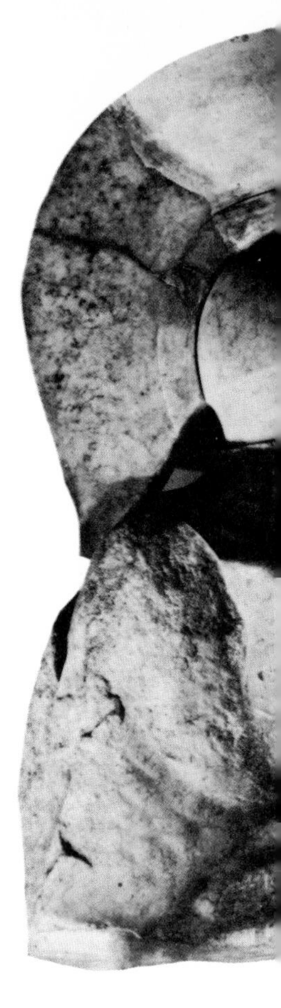

Leonidas
King of Sparta and Commander-in-chief of the Greek forces
As a true Spartan Leonidas could hardly have been more unlike the self-indulgent Xerxes. The Spartans were not the most attractive of peoples, and it is difficult to find in their society many redeeming features other than their indisputable courage and devotion to duty. Their daily lives were austere in the extreme. Physical perfection was of great importance, and young men and women exercised naked in the public arenas; all adult males ate apart in communal messes. But despite a gloomy and narrow outlook the Spartans were not without guile or political sense. Although very little is known of Leonidas's personal life, at Thermopylae he was to embody fully the ideals of his parent city, whose inflexible spirit is well represented by the instruction with which Spartan mothers sent their sons off to war; as each son was given his shield he was told to return 'with it or upon it'.

The Greeks' main line of defence was sited at one of the narrowest points in the pass, known as the Middle Gate; there Leonidas ordered a ruined wall to be rebuilt and on the morning of the battle placed his front line some distance ahead of the wall to receive the Persian assault.

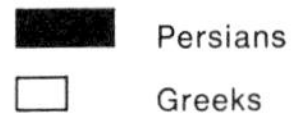
Persians
Greeks

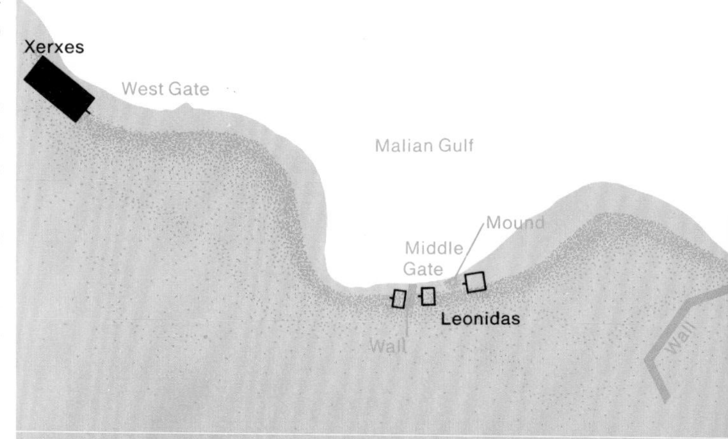

MAP 2 THE DEFENCE OF THE PASS

Xerxes I
King of Persia and Commander-in-chief of the Persian Army
Xerxes I, the 'Great King' (*c.* 519–*c.* 465 BC) was an absolute monarch. In character he was self-indulgent, not so much from a great love of luxury as from being so accustomed to the sybaritic life that he blandly accepted it as his proper due. As an administrator and organizer, however, he had much to commend him. Into many fields, notably those of communications and the Persian civil service he had introduced a degree of efficiency rare for those times; even so, his monetary policy was abysmally primitive, being chiefly governed by avarice. An impulsive and cruel man, Xerxes like many Persian rulers before him had been brought up to regard the lives of his subjects as his to control. Surprisingly perhaps, he did not shrink from the realization that he himself had no great talent as a soldier; in consequence he readily delegated tactical command to his subordinates.

There was little that the Greeks could do but await the tempest which sooner or later was inevitably to break. Across the pass at a point called 'the Middle Gate' stretched a ruined wall, which they rebuilt. It was in front of this rudimentary fortification that Xerxes' astonished scouts saw the Greeks strolling casually about, exercising and probably, in the case of the Spartans, dressing and combing their traditionally long hair.

The Persian king was in something of a dilemma. His fleet was not at hand and a mighty storm was raging over the sea; he seems to have decided that it would do no harm and not a little good to try to gain time – after all, his *was* a most formidable army, and the Greeks *might* just be apprehensive enough to accept any reasonable terms to abandon the pass. Accordingly he sent a herald to the Greeks offering them a free passage home if they allowed him through the pass. This offer, though causing some little debate among the recipients, was nevertheless firmly rejected and once more Xerxes was flung back upon his own initiative. His problems were increasing, not least because to feed and supply his army was a massive undertaking; and at the same time the veriest tyro in the art of war could see that it would be a considerable task to force a really determined opposition out of the pass.

The Course of the Battle

Three days passed while Xerxes debated the problem with his advisers. Finally his mind was made up by the news that his fleet had been badly battered by the storm just diminishing and that at least two days would be needed to carry out repairs. Xerxes decided there and then to make an all-out attack on the Greek position. A frontal assault was chosen because extensive probing by patrols had failed to uncover a route through the mountains by which the Greeks could be turned. It was 18 August 480 BC.

Orders went out to the Persian regiments and they began to move forward. Initially the Greeks formed up at some distance – possibly as much as 200 yards – in front of their stone wall, where they presented a further wall of raised shields to the advancing enemy.

Leonidas had detailed his Spartans, his *corps d'élite*, to receive the first onslaught, and they duly closed ranks, their great shields overlapping in the tightest formation that left them room to use their spears. With the line probably formed at the narrowest point of the defile, it might have numbered no more than fifty per rank, thus making the Spartans (some 300 strong) six men deep. On came the first wave of attackers, the initial shock being provided by a brigade of Medes, who together with the Persians were the dominant races of the Empire. They were men of great fighting ability; each was armed with a spear – shorter by a foot or more than those of the Greeks – and a light wicker shield.

Rank after rank of Medes rushed forward and in a matter of seconds the Spartan front was heaped with enemy dead and wounded. The line of bright bronze shields was unbreakable, and the glittering spearpoints of the Spartans darted venomously forth, thrusting through the robes and light armour of the enemy. The fight raged on, fresh men moving up to take the place of the slain, leaping over the hundreds of bodies that already lay in front of the Greeks. The clash of weapons, the yells and grunts of men in mortal combat and the cries of the wounded echoed weirdly from the high cliffs and across the waters of the gulf. Again and again the Medes returned to the attack, each time losing large numbers of men, until finally they gave way and had to be withdrawn into a reserve position. The Greeks were given a short breathing space before another assault-wave was dashed against the pass, this time by fresh Scythians and Cissian tribesmen, who attacked ferociously but made no impression on the Greek wall.

Xerxes was beside himself with fury and astonishment and although daylight was rapidly fading he decided that the business must be finished forthwith. He summoned Hydarnes, commander of the Immortals, and ordered him to sweep away the presumptuous Greeks with his guardsmen. The Immortals presented a daunting spectacle to the defenders of the pass – big, dark-skinned and black-bearded men, with flowing, multi-coloured robes and long shields. Although battered and exhausted the Greeks gripped their spears and shields and braced themselves.

The fighting was now of unparalleled desperation with the Greeks under the severest pressure. But the moment of peril passed – discipline, technique and superior weaponry again gave them the advantage. First one robed figure ran back to the rear, then another, followed by small groups until the entire assault-force of Immortals was fleeing in hopeless confusion. Thus ended the first day.

At first light the following morning Xerxes once more hurled his legions at the pass. This time he assembled a picked force, promised its members untold wealth if they were successful, threatened them if they were to fail and ordered them into battle. Again the Spartans and the other contingents relieved each other in the line, as they had done towards the end of the previous day, the men of each city state taking their share of the fighting. As attack followed attack, heaps of dead were strewn across the narrow pass; and on the Greek side, too, losses began to mount. Throughout the day the pass rang with the clash of weapons and the cries of men, and at length Xerxes was again forced to suspend operations. It was becoming increasingly difficult to force the Persian troops across the welter of dead and dying who now covered the ground. The situation, as far as the Great King was concerned, had reached stalemate; neither he nor his numerous generals could think what to do next.

At this stage there came upon the scene a man who was to relieve Xerxes of many of his difficulties. He was a Greek named Ephialtes. On being granted speech with the Persian King, he assured him that what his men had sought

top The Spartans were deputed to meet Xerxes' initial onslaught – supplied by a brigade of Medes – and they formed up in close order across the pass, their long spears poised. **above** Xerxes' infantry, right, surges towards the Greek lines. **right** Despite their brave defence of the main position, the Greeks found themselves outflanked on the morning of the third day by Xerxes' élite troops, the Immortals. As the end approached and the last Greek survivors were totally surrounded, they made their last stand on a low mound situated to the rear of the wall, where they died to a man.

in vain really did exist – there *was* a track through the mountains by which the Spartans' impregnable position could be outflanked and attacked from the rear. Xerxes was overjoyed and sent again for Hydarnes, who had commanded the Immortals in their disastrous attack the previous day. Hydarnes seized this opportunity to restore his status; preparations were quickly made and as night began to fall he set off with his Immortals. The traitor Ephialtes guided the column.

Throughout the night the Immortals half marched, half climbed through the narrow gaps indicated to them by their guide. They moved through boulder-strewn gorges and across rugged screes until, as first light appeared faintly in the east, they were marching rapidly along the mountain summit prior to making a swing northwards that would bring them down to a point behind the Greek position. As they descended the slopes they met the Greek flanking force, the thousand Phocians who had been entrusted with guarding this vulnerable approach. Alas, several undisturbed days had taken their toll of military vigilance, and the attack of the Immortals was sudden, unexpected and devastating. Unslinging their powerful bows, the Persians sent a volley of arrows whirring into the ranks of the Greeks. The Phocians were not of the stern stuff of their comrades in the pass itself and almost at once they scattered in confusion. For the Persians time was now a vitally important factor and Hydarnes drove his Immortals forward.

Total surprise was not to be effected, however, on the Greek army below. Leonidas had already been informed of the movement of the Immortals. During the night deserters, apparently conscripted Greeks from Ionian cities on the coast of Asia Minor, had crept out from the Persian camp and made their way into the pass, where they had been promptly seized by Greek outposts and brought to Leonidas. They had told him about the Greek traitor, the muster of the Immortals and their march into the mountains.

The Spartan king had seen the truth with terrible clarity, realizing that the guard he had stationed to protect the hidden track was hopelessly inadequate. There had been no time to send reinforcements to the danger point and shortly after daybreak messengers came racing down the mountain with news of the Phocians' flight and of the advance of the Persians towards the rear of the Greek position. Leonidas and his men prepared to be attacked from two sides. It was not long before they could see Xerxes' army forming up in front of the pass.

As soon as it had been confirmed that the Immortals were marching towards the rear of his position, Leonidas assembled the commanders of the various contingents making up his army. This was not an ordinary council of war – ordinarily the recourse of a weak and indecisive general. The Spartan king already knew what he and the remnant of his 300 were going to do, but he had to know what was in the minds of the leaders of the Thebans, the Locrians and the other contingents, for they were only formally under his command and he had no real authority over them. What took place at the council must remain a mystery, but afterwards several of the state units marched off to safety, leaving only the Thebans and the Thespians to support the now much diminished Spartans. The force remaining can hardly have amounted to 1,000 men; certainly there

above A Greek hoplite uses his heavy spear to dispatch a fallen adversary.

The map shows the secret route through the mountains taken by the Persian Immortals, who were led on their mission by the traitor Ephialtes. The Immortals marched through the night and in the morning they overran the Phocians stationed to guard the path and advanced without further hindrance on the rear of the main Greek army.

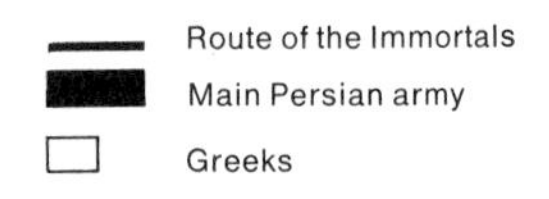

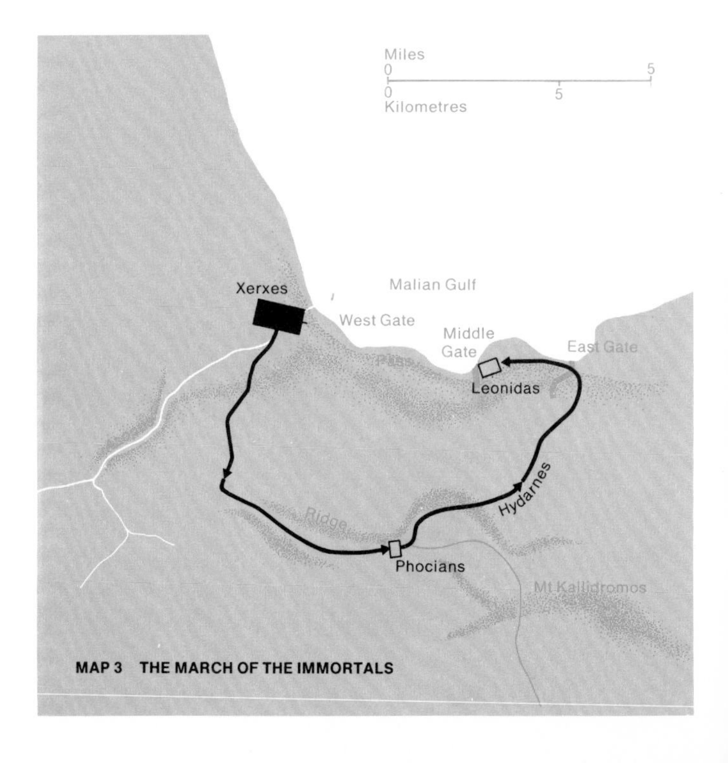

MAP 3 THE MARCH OF THE IMMORTALS

had been severe losses in the first two days of bitter fighting.

In these impossible conditions the defenders looked to their weapons, adjusted their armour and took up their positions. This time Leonidas advanced his line some considerable distance into a wider section of the pass; this gave a broader frontage for his hoplites, allowing more of them to meet their attackers at once and so cause a maximum of casualties. There they waited until late in the morning. When the Persians attacked they did so with great determination, well aware of course that their comrades would soon charge upon the Greeks from the rear. Wave after wave flooded up to the wall of shields then ebbed away; but each surge reduced by a small amount the number of Greeks still able to fight. At last the gallant Leonidas himself fell before an avalanche of Persians. The Spartans retrieved his body after some particularly savage fighting and dragged it back to their lines.

Still the remnant fought on, fewer and fewer in number, until the dread word came from the rear. The Immortals were arriving and they were totally surrounded. Steadily and deliberately the Greeks closed their ranks and marched back to a low mound which lay just behind the wall, their numbers by this time shrunk to a handful of men. There, upon the hillock, they fell to a man with their faces to the enemy, showered by javelins and arrows, until only the dead were left to mark the defence of Thermopylae.

Aftermath and Conclusion

That was the epic fight of Leonidas and his men. For the Greeks it was a defeat, albeit a glorious one. The outcome of Thermopylae sent a thrill of pride pulsing through the land, bringing a fresh sense of unity to the dissident Greek states. But much had yet to be endured, for the Persian hordes poured forward across Greece, burning, killing and looting. Athens, which had been evacuated save for a garrison on the Acropolis, was entered and the fortress itself stormed on about 5 September. The garrison and those who had sought refuge there were slaughtered. In the meantime, however, the Greek fleet was still an active force, based on the island of Salamis, whither many Athenians had fled for safety. After days of manoeuvre, the two fleets met in the Salamis Channel on 23 September 480 BC, and on that occasion the superior tactics and fighting power of the Greeks proved too much for Xerxes' vaunted navy. After a battle lasting upwards of eight hours the Persians were totally defeated.

Salamis was a shattering reversal for Xerxes. He was forced to consider retreat almost for the first time, and his thoughts flew to his Hellespont bridges, now painfully vulnerable to attack. A few days after the Battle of Salamis, he retraced his steps to the north and with much of his huge army retreated over the bridges into Asia Minor. Behind him he left a strong force in northern Greece under the command of Mardonius.

The remainder of the story may be briefly told. Athens was reoccupied by the Persians and again evacuated by them by the beginning of the year 479 BC. Mardonius was finally brought to battle by a Greek army, again under a Spartan king, this time Pausanias. At approximately the end of August Mardonius was defeated and killed at the Battle of Plataea – as significant a land battle as Salamis had been on the sea. After Plataea the Persians gave up their plans to add a part of southern Europe to their vast empire, and retreated for the last time into Asia. It was to be left to Alexander the Great (356-323 BC), almost 150 years later, to complete the restoration of Greek martial pride when he swept across Persia and destroyed the age-old enemy's power at the Battle of Arbela (331 BC), after which he penetrated still further eastward, reaching northern India by 326 BC. But for all the greatness of Alexander's achievements the seeds of Greek unity, so vital to her future prosperity, were undoubtedly sown by the men who fought and died at Thermopylae. The epitaph to Leonidas and his soldiers is brief and direct:

Tell them in Lacedaemon [Sparta], passer-by,
That here obedient to their word, we lie.

Philip Warner at

AGINCOURT

Cut off en route for Calais by a much larger French force of 25,000 led by Charles d'Albret, the Constable of France, Henry V of England brilliantly deployed his force of 5,000 archers and 1,000 men-at-arms along a narrow front flanked by impenetrable woods. The French army, compressed into the same narrow space, was slaughtered from above by the deadly arrows of the English longbowmen, and Henry brought his Normandy campaign to the triumphant political climax he had sought.

1415

English men-at-arms defend themselves during the first counter-attack of the French cavalry.

The Background to the Battle

Extraordinary though it may seem, the remoter causes of the Battle of Agincourt, which was fought on 25 October 1415, go back as far as 14 October 1066. By the Battle of Hastings, which took place on that date, William of Normandy conquered Harold of England, and thereby took the English throne as William I. However, as Duke of Normandy, the new King of England was still vassal (subordinate) to the King of France. This feudal allegiance was made more complicated by Henry of Anjou, William's grandson and later Henry II of England. In 1152 he married the wealthy Eleanor of Aquitaine and acquired vast domains in south-west France.

From then on the English kings were potentially, if not always in reality, as wealthy as their French nominal overlords. But over the years they found their French territories difficult to control, and these dwindled despite a brief recovery under Edward III, whose brilliant campaigns included the battles of Crécy (1346) and Poitiers (1356). Even so, two years before Edward III's death, English possessions in France were limited to Calais and a few small territories elsewhere.

But when in 1413 Henry V came to the throne, young, devout, and as able a soldier as Edward III, his grandfather, a new era began. Henry claimed the French throne as being his by right through his descent from Isabella of France, wife of his great-great-grandfather, Edward II. Her claim had in fact been set on one side long before, and Henry's would now have been dismissed as a ludicrous fantasy but for two important factors. The first was that the reigning King of France at Henry V's accession was Charles VI, who was frequently insane and never competent even at the best of times. Furthermore, the Dukes of Burgundy and Orleans, who alone could have stabilized France, were bitter and active rivals. Henry V was able to appear friendly to both, though in fact he favoured Burgundy.

Last but by no means least in the chain of causes which sent Henry on his road to French conquest was his unshakeable conviction that it was his right and duty to defeat the French in battle and rule over them. This great conviction, allied to Henry's powers of enthusiasm and limitless energy, soon inspired others. Parliament voted a subsidy for the coming war; Henry increased it by securing loans (most of which were eventually repaid). Men were called up or volunteered for service; Henry ensured that every detail of necessary equipment would be available for their use. This meant not merely guns, bows, arrows and armour, but masons and miners, cordwainers (leather workers) and carpenters, bakers and butchers – and spares for everything. This was no army planning to live off the country, but a new-style task-force that could go where it wished. At least, that was the intention. Nevertheless, although a confident, trained, well-equipped and homogeneous force, it was

Delays in the taking of Harfleur forced King Henry V of England to abandon the original aim of his French campaign of 1415 – which was to advance directly on Paris. Instead he set off with 6,000 men on a march of conquest across northern France towards Calais, which at that time was English-held territory. But at the Blanche-Taque ford across the River Somme Henry's way was blocked by a French advance guard, led by Marshal Boucicaut, and he was forced eastwards as far as Voyennes before a crossing could be made. Now pursued by a combined French force of some 25,000 men under Charles d'Albret, the Constable of France, Henry was eventually cut off near Agincourt, where he prepared to do battle.

English

French advance-guard

French main body (joins the advance guard at Amiens)

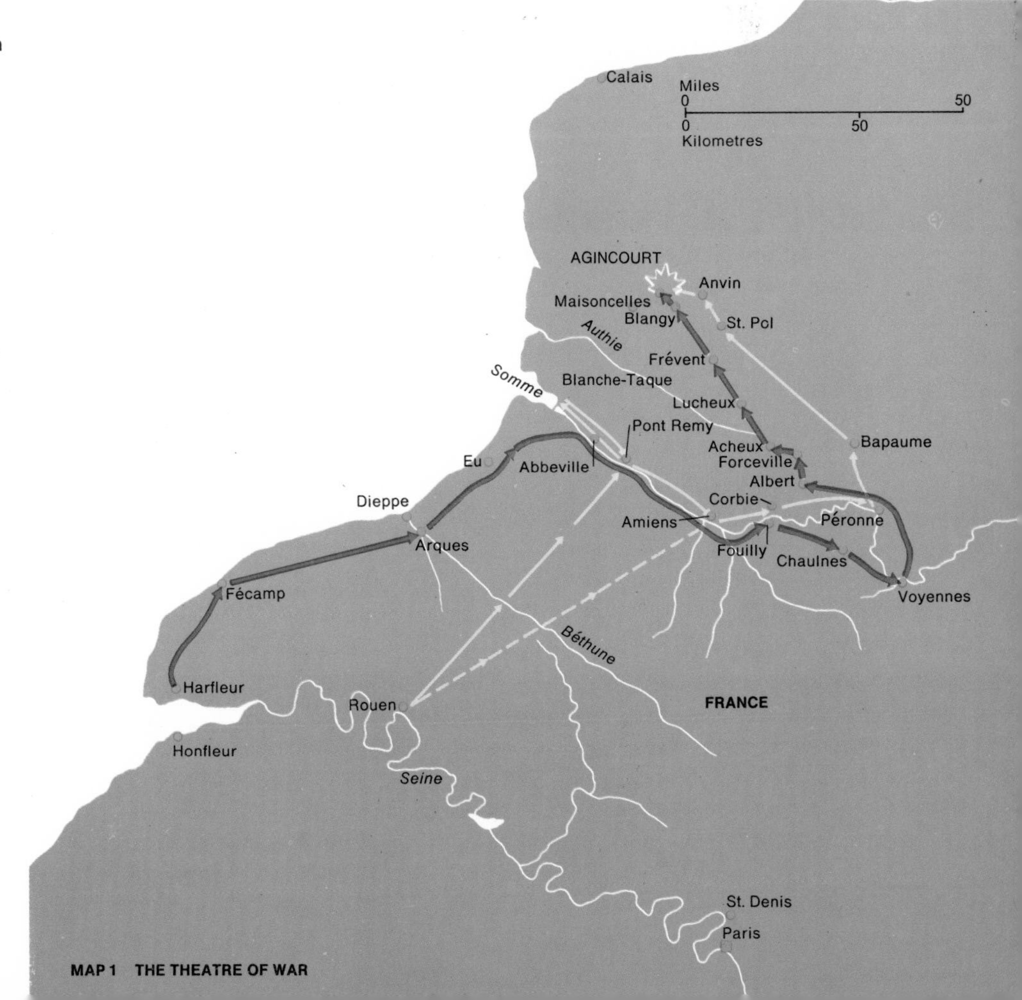

MAP 1 THE THEATRE OF WAR

right and below Henry V's helmet and shield, from his chantry in Westminster Abbey, London.

small in relation to the possible opposition, and Henry knew as well as anyone that success was not inevitable.

On 11 August 1415 the great invasion fleet set off. The strategic plan was very simple, although it took the French by surprise. Past experience suggested the invasion would begin at Boulogne, and the French were in partial readiness there; Henry, however, had a different plan, and a bold one. Maintaining complete secrecy, he set off for Harfleur at the mouth of the Seine. Harfleur was an important port before silting affected it, and it was a superb base for an invasion – if it could be easily captured. It was undoubtedly enterprising to stake everything on a surprise attack, far removed from the firm foothold of Calais – but it was also risky. Harfleur was, however, much nearer to Paris, the ultimate destination, and if all went well Henry could slip past the assembling French armies and be at the capital before they could prevent him. It was a calculated risk, but not quite as nicely calculated as it should have been.

When the English landed on 14 August Henry probably estimated that three days would be sufficient to unload and that he would need a week or so to take the town. The first part went according to plan but the second calculation went badly awry. Harfleur had formidable defences with wide ditches, strong towers, thick walls and jutting emplacements which gave the defenders an excellent field of fire. Usually sieges were won because the miners dug under the walls, but on this occasion the countermining was so effective that the efforts of the English had to be abandoned. Instead, the guns were pushed forward and the vital walls were eventually reduced to rubble; the siege of Harfleur was, incidentally, one of the earliest successes for artillery in the history of warfare.

However, when Henry received the surrender on 22 September his strategy had gone to the winds. In those vital weeks two important events had occurred: one was the reduction of his invasion force by dysentery and other debilitating or fatal diseases; at least 2,000 had been killed, had otherwise died or been invalided, and a further thousand had had to be left in Harfleur as a garrison force. From an original strength of nearly 10,000, Henry was now down to little over 6,000. The other important factor was that the French had had sufficient time to gather a formidable army.

This was the moment at which Henry thought it wise to issue a personal challenge to the Dauphin (the French King's eldest son) to meet him in single combat and so determine who should ascend the throne of France on the King's death. As everyone expected, the challenge was declined.

Tempting though it was to press on to Paris, in the hope of still achieving surprise with his mobile, though depleted, army, Henry reluctantly had to abandon this plan. With 6,000 men, of whom 5,000 were archers, he would inevitably be outnumbered and might be outmanoeuvred too. Harfleur had taught him a memorable lesson. But equally

left Early fifteenth-century pole-axes. These versatile weapons were made with a spike for thrusting and an axe-head and a hammer-head for downward blows.
left Siege warfare in the fifteenth century. The main body of the attacking force is drawn up ready to charge forward once the walls are breached by cannon-fire.

right The rival armies before the Battle of Agincourt, as seen from a position behind the English centre. This view illustrates the use of the herce, a hollow wedge formation that gave the archers a greater field of fire, enabling them to attack the enemy's flanks. The English line contained four herces and further parties of archers on either flank.

firmly he rejected the advice of his War Council, which suggested a rapid return to England. Doubtless, like all cautious policies it had much to recommend it. However, cautious policies do not win campaigns; often they contribute to losing them.

It was unthinkable that Henry should tamely retire, but it was also impossible for him to take Paris. Even so, if his claim to the French throne was not to turn into a poor joke he had to produce something more spectacular on this occasion than the capture of one port. He decided to show he could march through France at will – at least as far as Calais. It is, of course, a disturbing and humiliating matter to have an enemy army marching across your territory, and Henry doubtless thought that that was exactly what was required to lower French morale. There were, of course, precedents in this activity, which was known as *chevauchée* – literally, an expedition on horseback. John of Gaunt had done it, and the Black Prince had done it. It showed a suitable disregard for the feelings of the inhabitants.

But, there were difficulties. Henry would have to cross a number of rivers on his march to Calais, and the Somme was fordable in only a few places. Furthermore, if he made any mistakes he would be confronted with the vast French army which Charles d'Albret, the Constable of France and Commander-in-chief of the army in the absence of the King, was said to be assembling at Rouen. Nevertheless, it would be a magnificent gesture, and it would convince those who had already lent him money that they could well now lend him a little more.

The army which left Harfleur on 7 October was light and fast. It had no guns and a very small baggage train. It carried its own rations, partly because Henry was opposed to plunder and partly because the countryside was thought to have been stripped already by the French.

Events did not go easily. There was occasional local opposition, some of it, as at Eu, strong enough to be disconcerting. Dysentery was a dangerous nuisance, but all seemed reasonably well with the English until they were within a few miles of the Blanche-Taque ford on the Somme. There, to their dismay, they learned that the ford was staked to prevent their passage, and guarded by 6,000 men under Marshal Boucicaut.

Henry's choice of route had proved his critical mistake; it remained to be seen whether the French could redeem him by making an even bigger one themselves. But, as he turned east to look for another crossing, the French were steadily assembling a force approaching 60,000 men – evidently forgetting their personal rivalries in the national emergency. The future looked ominous for Henry.

As the English army moved along the river, searching for an undefended crossing-place, food began to run short. The French had left nothing; and nuts, water and a meagre ration of dried beef are not much of a diet for a long march. Once the English soldiers found wine, but before they could besot themselves completely with it Henry had

right The armies faced each other on newly ploughed land flanked on both sides by woods – those surrounding the castle and village of Agincourt on one side and the Tramcourt woods on the other. The open space between was 950 yards wide, except at the French end where it widened to 1,200 yards. The armies initially formed up 1,000 yards apart, then the English advanced to within 200 yards. The ground was wet and extremely soft.

ENGLISH

Men-at-arms

Archers

1 Camoys
2 Henry V
3 York

FRENCH

Men-at-arms (mounted)

(dismounted)

Archers

1 Vendôme
2 Bourbon
3 D'Albret
4 Boucicaut
5 Orléans
6 Eu
7 Bourdon
8 Alençon
9 Bar
10 Brabant
11 Lammartin
12 Marle
13 Fauquemberghes

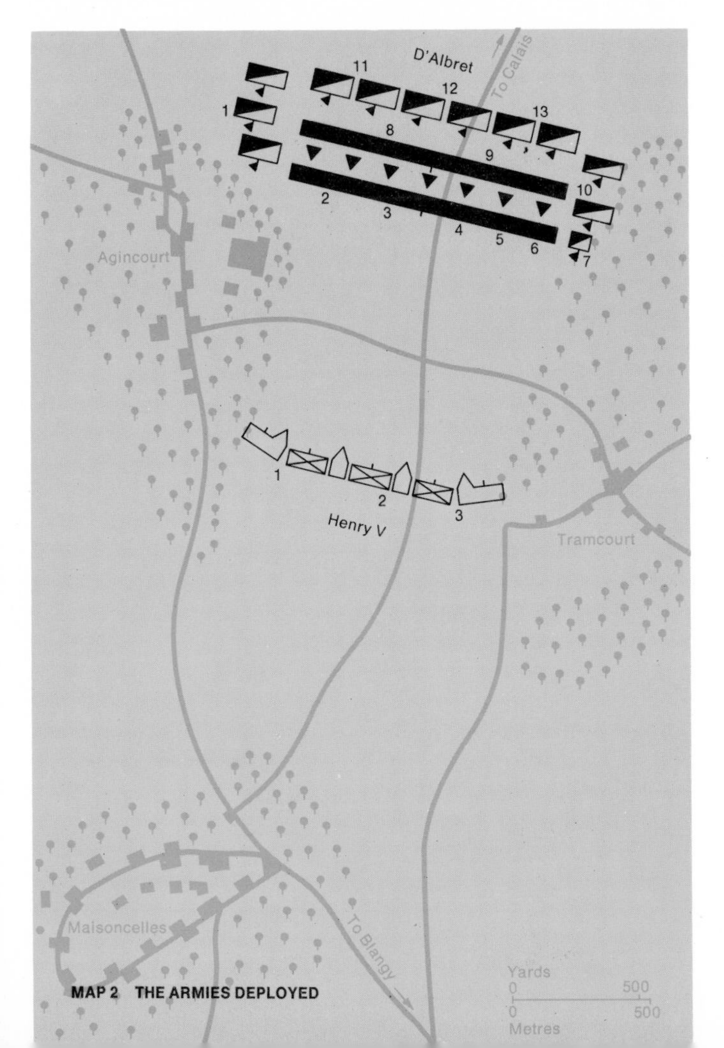

MAP 2 THE ARMIES DEPLOYED

Henry V

King of England and Commander-in-chief of the English Army

Henry's Normandy campaign was fired by the King's personal conviction that the French crown was his by right. Under Henry II England had owned vast territories in France which the French had since repossessed. He also saw it as vital that the French, too long the hereditary enemy, should be subdued once and for all: since the turn of the century French raiders had appeared in the English Channel and had made several destructive landings. The French business, Henry told his Parliament, needed an urgent solution.

Henry V, born in 1387, came to the throne in 1413 and died, on a campaign, in 1422. Most of his life was spent in warfare. He was a good strategist, enterprising but flexible, and a master of tactical appreciation even in the hottest phase of a battle. He was quiet and deliberate in manner, but was considered to be harsh and ruthless to his political and religious opponents; he was an adamant disciplinarian in the field.

Charles d'Albret

Constable of France and Commander-in-chief of the French Army

Charles d'Albret was Constable of France, an office which made him effective commander-in-chief of the army in the absence of the enfeebled King Charles VI, seldom sane, and rarely competent to command his forces. By a further, though fairly unimportant, complication, d'Albret was not in fact titular commander: for reasons of protocol this honour was accorded to the Dauphin, the King's eldest son, a person of little note and only nineteen years old.

D'Albret had a difficult command from the beginning. Not only was he himself regarded with some suspicion by the French court, he had at the same time to try to weld together a fighting machine from various unco-ordinated parts. Possibly d'Albret under-rated Henry's army and his archers, but as he was killed in the battle he had no chance to justify his handling of his troops.

below After the deadly opening fire of the English longbowmen the French counter-attacked with their flank cavalry. They failed seriously to penetrate the English line of stakes, however, and soon were charging back to seek refuge behind their own tightly packed front line. Disaster followed when the close ranks of heavily armoured men-at-arms were unable to manoeuvre quickly enough to let the cavalry through; many were knocked down, trampled and suffocated in the soft mud.

ENGLISH

Men-at-arms

Archers

1 Camoys
2 Henry V
3 York

FRENCH

Men-at-arms (mounted)

(dismounted)

Archers

1 Vendôme
2 Bourbon
3 D'Albret
4 Boucicaut
5 Orléans
6 Eu
7 Bourdon
8 Alençon
9 Bar
10 Brabant
11 Lammartin
12 Marle
13 Fauquemberghes

ordered them on. It seemed harsh – as they were clearly doomed, and the end could not be far distant. But Henry was having no nonsense – no nonsense at all. When a soldier stole a pyx (a holy vessel) thinking it was gold, Henry had him hanged and made the army march by his suspended body. And when, further on, near Fouilly, they were surprised by a sudden foray by some French knights – soon driven off, but dangerous for a while – Henry ordered his soldiers to cut six-foot-long stakes and carry them. They would be worth their weight in gold later – for repelling cavalry attacks – but for the moment they were a tiresome burden for weary men. The soldiers grumbled, but not within earshot of the King.

Eventually, by cutting across a loop in the river, Henry outpaced the French on the far bank, and reached an unguarded ford at Voyennes. It was not an easy crossing but they managed it without mishap. Now with renewed optimism they turned their faces towards Calais, still 100 miles away but without serious natural or human obstacles on the route. Or so they thought.

The French, however, had not been idle. By 19 October they had moved directly across the path which Henry planned to take; they were, in fact, at Péronne, immediately between Henry and Calais. There they hesitated and reorganized before sending a challenge to the English king. In the best convention of medieval warfare the French requested Henry to cease his march to Calais but, if he

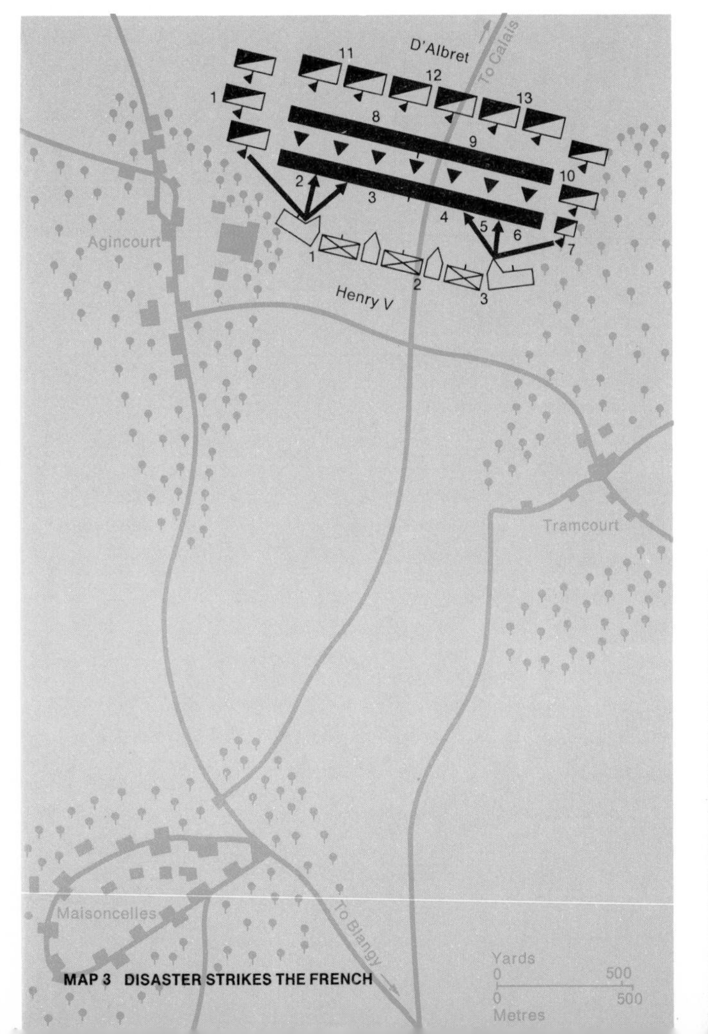

MAP 3 DISASTER STRIKES THE FRENCH

wished to persist, invited him to choose a suitable place at which to fight. Henry declined to choose his ground and hoped that in consequence the French would attack him in his present position. When they showed no inclination to do so he moved forward along the line Albert–Acheux–Lucheux–Frévent–Blangy. His army was now marching at an average speed of seventeen miles a day. It was hard going; they were short of food, and the stakes and battle equipment seemed heavier as time went on.

Just as the English scouts approached Blangy, they saw the French army. It was a terrifying sight. There seemed to be no limit to them as they totally blocked the English line of advance. Clearly it was all over, bar the slaughter – and the small English army felt sure, with such a disparity in numbers, who would be slaughtered. On 24 October 1415 the two armies settled down for the night, the French jubilantly, speculating on their easy victory the next day, the English resignedly, resolved to make the French pay for every sword cut. It was raining, as it had been for days. It is possible for men to sleep in such conditions, with no cover, on wet earth, if they are tired enough, and hard enough, and Henry's men slept. Their bowstrings were well saturated with beeswax to keep out the rain, and each man had a spare string coiled inside his cap. The wooden shaft of the arrow (called the 'stele') was coated with 'virido greco', a copper solution which made it look like greenwood; this kept out the damp.

A mounted French knight charges the English *chevaux-de-frise* – a line of six-foot stakes hammered into the ground to protect the archers from cavalry attacks.

The battlefield which Henry saw before him the following day could not have suited him better if he had chosen it himself. On his left were woods surrounding the castle and village of Agincourt – then known by that name, though subsequently it was changed to Azincourt. On the right were the Tramcourt woods. The open space between was 950 yards wide, except at the French end where it widened to 1,200 yards. The armies confronted each other 1,000 yards apart. The ground sloped slightly down to the middle giving each army a good view of the other, and – as Colonel A. H. Burne first noted after writers had been describing the battlefield for five and a half centuries – also fell away appreciably to the flanks. The ground had been ploughed recently and had been sown with wheat. It was wet, slippery and soft.

The Rival Armies

The armies deployed for battle. Their comparative strengths were approximately as follows:

English		**French**	
Archers	5,000	Cavalry	1,200
Men-at-arms	1,000	Men-at-arms and archers	23,800
Total	6,000	*Total*	25,000

There may have been slightly fewer English in the line because the weakest men had been left to guard the baggage. The English were drawn up in three 'main' divisions of men-at-arms, these being the foundation of the army; but as there were only 1,000 men-at-arms to 5,000 archers, the formation must have looked strangely unbalanced to the French who faced it.

Between the divisions were 'herces' of archers; these took the form of hollow wedges which projected, point forward, a short distance ahead of the front line. From the herce the archers could fire into the flanks of the advancing formations; the front line often knelt and so could take a heavy toll on those of the enemy who raised their shields to protect themselves from higher-flighted arrows. A herce often consisted of 1,000 men advancing as an isosceles triangle, their sides 200 yards long; at Agincourt they were probably slightly fewer, for there were four herces in all, and projecting beyond the outside flank herces were other parties of archers. The archers were, in all probability, divided equally between those in the herces and those on the wings – although there is a school of thought which believes that two-thirds of the numbers available were used in the wing formations. Certainly this was a battle in which flanking fire was used to the utmost effect. It may be that Henry anticipated a French advance through his middle, whereupon he planned to turn the tables by escaping through the woods and re-forming after he had inflicted heavy casualties, and then to resume his march to Calais. Henry himself commanded the centre

above The French flank cavalry encounters the English *chevaux-de-frise*. The main French lines, though reeling under the early fire of the English longbows, are so far uncommitted. **right** The confused scene as the English archers advance to retrieve their arrows (each carried a quiver of twenty-four and some spares) and find the French front line threshing helplessly in the mud in complete disarray. **far right** A moment from the grim episode that occurred when Henry V, wrongly informed that a second and equally large French force was approaching, ordered his men to slaughter their prisoners.

division, the Duke of York was on his right and Lord Camoys was on his left. The formation comfortably and exactly occupied the frontage of 950 yards.

The French, with vastly greater numbers, had a very different formation. Their force consisted almost entirely of knights and men-at-arms, although there were a few crossbowmen and longbowmen interspersed. As d'Albret had 25,000 men to deploy in a narrow frontage it was obvious that his divisions must be in tighter formation and much deeper.

The Course of the Battle

For four hours on that wet morning of 25 October 1415, the two armies confronted each other. In the medieval custom each was waiting for the other to make the first move and therefore leave an exploitable opening. Doubtless there were challenges and counter-challenges, but for all that it must have been an uneasy period. At 11 a.m. Henry decided to take the initiative. With the order 'Advance banners', the army moved forward. The battle had begun.

At 200 yards from the French line Henry gave the order to halt and fix stakes. With some relief the archers slipped them off their shoulders and hammered them into the ground with their 'mauls'. Every archer carried one of these leaden-headed mallets, which had a variety of uses ranging from hammering in a stake to cracking the armour of a fallen adversary to knife him through the opening. The archers then loosed off the first volleys of arrows.

The longbow was essentially an English weapon, though it was doubtless Welsh in its origins. Other European countries made little headway with it, yet every effort was made to encourage its use. A longbowman could send off six arrows a minute without exerting himself, and if pressed could manage twelve; during this time the crossbowman would send one or perhaps two. As the first shower fell on the French lines, some falling from the sky, some hissing through the air like snipers' bullets, the French shook and wilted. It was an expected move and they had a counter for it. Two flank parties – or at least those of them who could still control their horses, for more arrows buried themselves in the horses than in their riders – launched themselves onto the English line. As they reached it, in no slight confusion, they came on to the *chevaux-de-frise* or line of stakes set up by the archers. It was not an unqualified success for the English, for many of the stakes collapsed in the soft ground, but a number of French horsemen fell and were quickly despatched; the remainder wheeled and tried to recover through their own lines.

Had the French been in reasonably open formation they could have widened and let the cavalry through, but in those close-packed ranks such a manoeuvre was impossible. By the time the French had tried to deploy by advancing in column formation the damage had been done. Riderless or unmanageable horses raced back into the line knocking over heavily-armoured and thus clumsy men-at-arms and

An incident from the final, doomed charge of the French third line. In all the French lost 10,000 soldiers at Agincourt, many of their commanders, including Charles d'Albret, and a further 1,000 were taken prisoner; in contrast the English lost only a few hundred men.

knights. Once down in that soft mud, already being trampled into a liquid morass, the chances of their rising to their feet were negligible.

Still the arrows were coming down. Any attempts to rally the divisions in the rain, amid the screaming, the jostling and the confusion, were completely nullified by those pitiless flights of arrows coming down like dark birds from the sky. There was no avoiding them, but that did not prevent men and horses from plunging wildly as they received agonizing gashes from the long arrows, their sharp points greased for deeper penetration. Within five minutes the damage was done; but the arrows had gone.

The drawback to archery is that once the quiver-full of twenty-four arrows and some spares has been released, the archer has to receive more from the rear or recover those he has sent. Arrows were in those days torn out of the bodies of friend and foe alike with scant ceremony. There was no doubt where the arrows now were, and the archers surged forward to recover them. Soon, however, they were ignoring their need for arrows and settling for grimmer work.

The whole of the French front line was one struggling, helpless heap, so tightly packed that the most courageous could not raise an arm to defend himself, and the ground so slippery and treacherous that men were falling every moment, and being used as stepping-stones for others who would soon join them. Most were too exhausted to struggle, so encumbered were they with their heavy armour. Now the archers, many of whom were, in contrast, half-naked, were in their element. They dodged from one victim to another, hopping nimbly from one recumbent body to the next to deliver the crafty thrust that would add yet another to the mounting toll of French dead. The whole of the French front line had not merely disintegrated; it was being eliminated. The crossbowmen between the first and second French lines were unable to see a separate target, and it was only a matter of minutes before they too were driven forward by the relentless plodding advance of the second line. And as that line came forward, it too was overwhelmed, partly by its own men trying to fall back to a position where they could fight, partly by the triumphantly excited archers who were now clambering over the armoured mound of dead and dying like goats on crags. The second line, which also had its problems in advancing over several hundred yards of wet ploughland, gave even less fight than the first.

But the battle, even after this first appalling shock, was not over. The third French line was still uncommitted, and Henry kept a watchful eye on them – for they alone outnumbered his entire force. At the same time he allowed his men to take prisoners. This part of the battle was what most of them had come to France for and, provided it did not jeopardize the result, he would not stop it. It was a curiously methodical scene. Valuable prisoners were dragged out from the heap, unhelmeted and disarmed. Then they were claimed and sent towards the rear. The English would fight even harder to preserve their prizes than they would for their lives. But Henry kept control of his men. Ironically too, the English force was protected from a sudden French attack by a wall of French dead, and even cavalry could not move fast enough to be dangerous over that sodden field. In any event, the French third line had considerable doubts whether this was the time and place to attack. The casualties had almost all been on the French side; the English had merely lost a few hundred, even though they had included such eminent figures as the Duke of York and the Earl of Suffolk.

At that point there occurred an event which did not affect the result of the battle, poised though that still was, but which badly damaged Henry's reputation in the eyes

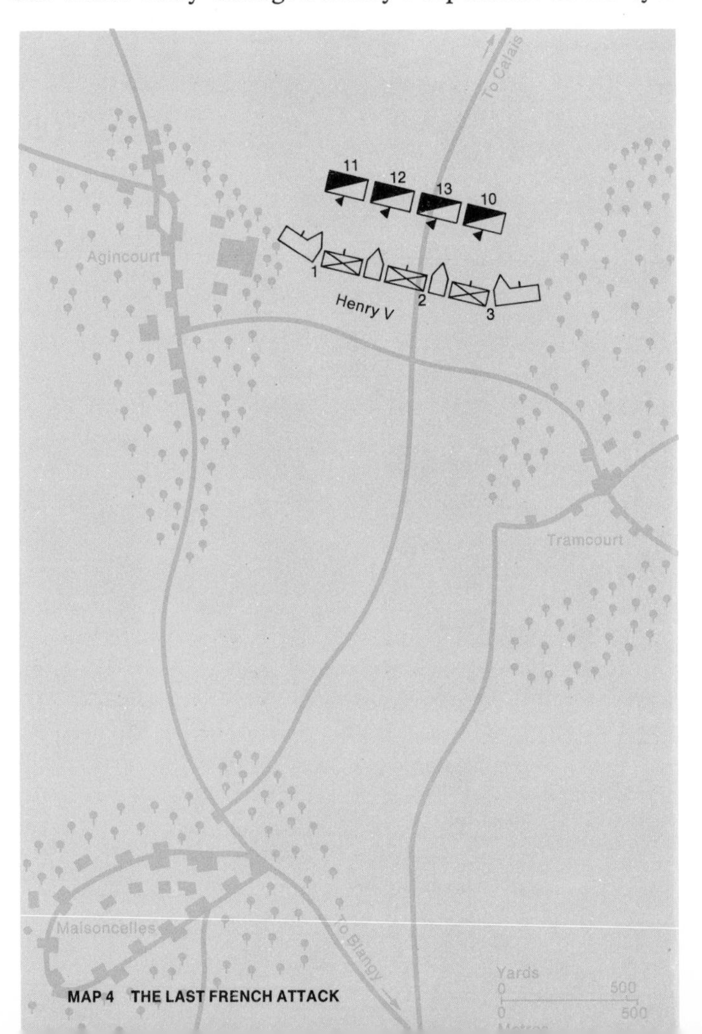

MAP 4 THE LAST FRENCH ATTACK

of posterity. The night before the battle he had been billeted in Maisoncelles, a mile behind the battlefield, and on taking the field had left his personal baggage there, including his crown and seals. When the battle began a local squire, anticipating an English defeat, decided, with a rabble of peasants, to help himself to the royal effects before the victorious French army should arrive. Henry's small baggage guard was outnumbered and overwhelmed in the raid, and they rushed forward with the story – exaggerated as most stories are when they come from those who have left a field of battle prematurely – that a huge French force was now about to descend on the English army.

Their story, though unlikely, was not impossible. Another 20,000 French could easily have been in the vicinity. Taking no chances, Henry ordered all prisoners to be killed. It was an unsavoury action, but as many of the prisoners were still in armour, and there were weapons lying around everywhere, there was little likelihood that they would remain prisoners for long if the English army was caught between this new force of unknown strength and the still uncommitted third line. Contemporary chroniclers – even French ones – found Henry's action both sensible and rational; some of the French even blamed their own side for bringing it about.

The English army began their cold-blooded slaughter so tardily and reluctantly that Henry had to despatch a special force to hurry on the work. But it was not humanitarianism which was holding up the killing; it was the thought of losing large sums of ransom money. Ironically, a few of the richest and noblest men were spared – for the benefit of the richest and noblest English. (Nobody would be surprised at that in medieval warfare.)

Then the news came through that it had all been a false alarm, and there was no new French force. The soldiery returned to their work of plundering or identifying the heaps of recumbent figures. Now at long last the French third line made a half-hearted attempt. Most of it had slipped quietly away, but those left made a vain gesture, including the Duke of Brabant, who in effect joined a lost battle just in time to be killed.

Almost at the end of the day the French third line – or what remained of it – made a half-hearted attempt at a charge, resulting in the loss of still more lives, including that of the Duke of Brabant, who in effect joined a lost battle just in time to be killed.

ENGLISH

Men-at-arms

Archers

1 Camoys
2 Henry V
3 York

FRENCH

Men-at-arms (mounted)

10 Brabant
11 Lammartin
12 Marle
13 Fauquemberghes

Aftermath and Conclusion

Apart from the appalling slaughter, which included not only 10,000 French soldiers but also Charles d'Albret, the Commander-in-chief, and the Dukes of Brabant, Bar and Alençon, the French also lost over a thousand prisoners, including Marshal Boucicaut, the Dukes of Orléans and Bourbon, and the Count of Eu. To many it must have seemed a form of national suicide, for the French had contributed largely to their own downfall by clinging to obsolete equipment and methods many years after they had been shown to be a dangerous liability.

So great was the destruction of men, materials and morale that Henry could have marched straight to Paris with negligible opposition. But whether he could have held the French crown if he had obtained it, or benefited anyone thereafter, is debatable. Instead he marched to Calais, and thence to a triumphant homecoming.

Agincourt was a curious battle in many ways. If the French had tried to lose it they could not have deployed their troops in a very different way. Possibly d'Albret felt that if the English advanced through a funnel 950 yards wide they could be slaughtered piecemeal. And the trees, even though they prevented him from using his own cavalry for flanking movements, would serve, he may also have thought, to prevent the English bowmen from deploying wide. (A bow can be used under many conditions, and even in the dark, but not through a canopy of trees.) Such ideas may have formed part of d'Albret's negative thinking, by which he ultimately permitted the massacre of his army.

Even so, it was touch and go until the last of the French third line left the field. If that line had gone in wholeheartedly – and it was still fresh, which was more than could be said for the English army – the result of Agincourt might have been very different. As it was, French morale after the battle was so depressed that the country's military leaders radically changed their approach to fighting in the open field; from then on they relied on delaying tactics and siegecraft.

Henry himself was no paper strategist: he believed in leading his troops into action and fighting with them. It was said of him that his prowess at Agincourt was worth that of ten ordinary men, and this may well have been true; among the few accounts that emerged from a very confused scene, there is one which refers to his rescue of his brother, the Duke of Gloucester, who had been wounded. Henry was undoubtedly fitter and more active than most of his own men and for that reason – as well as his mastery of tactics – he was a superb battle commander.

Peter Young at

EDGEHILL

In this first crucial test of strength in the English Civil War the army of Charles I faced a Parliamentarian force led by the Earl of Essex. The fight was long and bloody, a sternly waged clash of sword and pistol, cannon, pike and musket; at the end of the day when both sides, exhausted, broke off the struggle losses were even – but the moral advantage was with the King.

1642

A company of Royalist foot is armed and trained for war. In July 1642 King Charles I, having quit his capital and removed to York, issued commissions to those of his followers who felt they had the resources to raise regiments, troops or companies of men.

The Background to the Battle

Edgehill was not only the first great battle of the English Civil War, but the first important battle on English soil for more than three generations. The great majority of the officers and soldiers engaged had less than three months' service and were now in action for the first time. Drill, discipline and administration were all inevitably defective.

The English Civil War began in 1642 and lasted eventually for ten years. It arose from a struggle for power between the Long Parliament and the Royalists, the latter being led by King Charles I (1600–49) and his son the Prince of Wales, later to become King Charles II (1630–85). Its causes were not, however, entirely political. Had the political element not been exacerbated by religious quarrels there would have been no outbreak of war. The great majority of the population, then not more than five million, did not take up arms. It would indeed be very surprising to find that more than 150–200,000 men enlisted during the whole war, and most of these were garrison soldiers who seldom saw action in the field.

In religion the Royalists, also known as the Cavaliers, were generally Anglicans (Arminians) and Roman Catholics; the Parliamentarians, or Roundheads, were Presbyterians and Independents. But by 1644 the Presbyterians had come to detest the Independents rather more than they did the Cavaliers; still later, in 1660, the restoration of the monarchy (when Charles II came to the throne) was brought about by a coalition of Royalists and Presbyterians, who by then had become totally disenchanted with their 'fellow-travellers'.

Before the outbreak of war there had been long-standing opposition to the King's policies. In Parliament this opposition was led by a group known as the Five Members. In January 1642 matters reached a climax when King Charles attempted to arrest the Five Members. His orders were resisted and England drifted into war. The King quit London, his capital, and removed to York, where towards the end of July he issued commissions to those of his adherents who thought they had the resources to raise regiments, troops or companies of horse- and foot-soldiers.

The Parliament commissioned the Earl of Essex as Captain-General. With the armouries of the Tower of London and of Hull in its power it had little difficulty in supplying its levies. The financial resources of the City of London guaranteed that the cavalry should be well-mounted, the soldiers well-paid, the train of artillery well-found. But money alone cannot make soldiers, and experienced officers were lacking. In this respect the Royalists had the advantage, for the majority of English professional soldiers felt the innate loyalty of their caste towards the anointed King. The Cavaliers, dogged throughout the war by lack of ready money, could never have sustained the Royal cause for four years without the skill and courage of the professional soldiers who commanded their regiments and their armies. The Earl of Essex, on

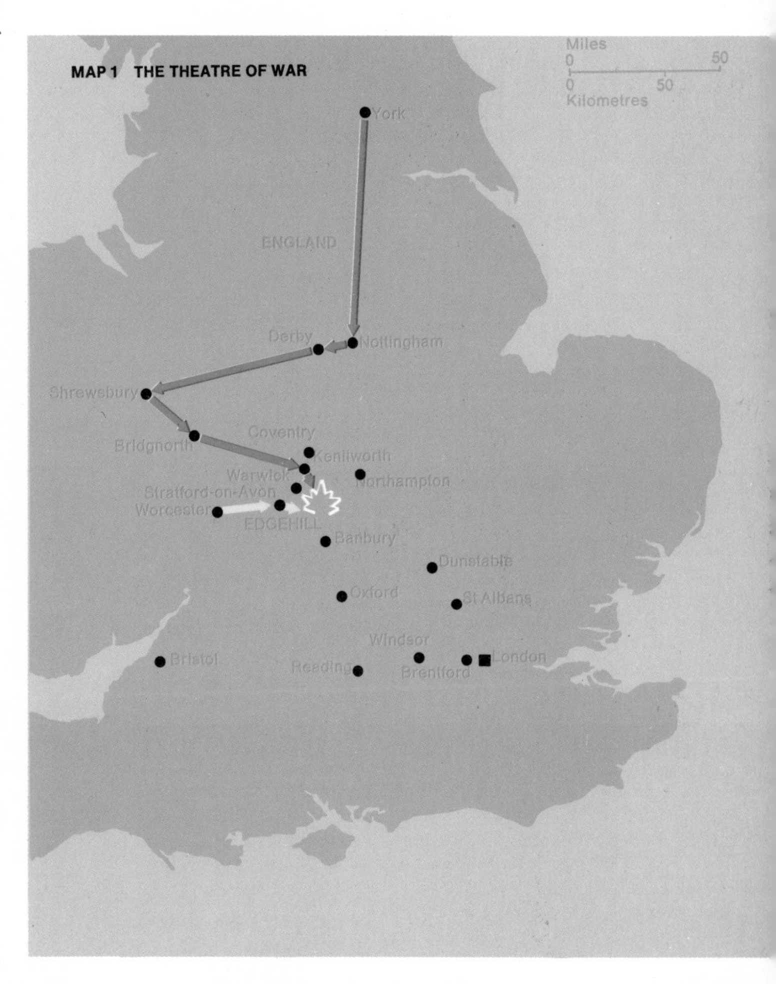

The routes to Edgehill taken by both armies are shown. On 22 August 1642 the King raised his standard at Nottingham. A minor victory at Powick Bridge, near Worcester, raised morale and on 12 October he began his march on London. The Battle of Edgehill was fought on 23 October.

Royalists

Parliamentarians

the other hand, was at best a lethargic commander, and a poor disciplinarian. His army was ready first but, being in rebellion, he was reluctant to strike the first blow; instead he waited for the King to take the initiative.

After raising his standard at Nottingham on 22 August 1642, Charles, his troops still seriously outnumbered, removed to Shrewsbury. The people of Wales and the Severn Valley were loyal to his cause and fresh levies came pouring in. A minor victory at Powick Bridge, near Worcester, on 23 September raised the morale of his cavalry, and on 12 October he began his march to London.

Essex had stationed himself at Worcester, which is not, of course, on the direct route from Shrewsbury to London. The Earl was an indifferent strategist, and it is easy to dismiss this as a mere blunder. In fact, however, he may not unreasonably have supposed that the King would open his campaign with an attempt on Bristol, then the second city in the kingdom, and a port where recruits, money, guns, ammunition and ships would all be forthcoming. Besides, many of the leaders of the community in Bristol were known to be Royalist in sympathy.

So, at that point, the King stole a march on the Earl, but this was no *blitzkreig*. Both armies were well satisfied if they moved ten miles a day. The roads were broad, unmetalled stretches of pothole and rut. The cannon, drawn tandem by heavy horses whose strength was ill deployed in the days before artillery horses were harnessed

below A cavalryman on the Parliamentarian side; like his opposite number in the Royalist army, he wears a buff leather coat, breast- and back-plates and a bridle gauntlet to guard his vulnerable left hand. His helmet is of the three-barred 'pot' type, also called a 'lobster-tail' pot after the jointed-plate construction of the tailpiece.

above A 9-pounder demi-culverin, dated 1580, mounted on a modern replica of the original gun carriage. The Royalist artillery had six big guns at Edgehill – of which two were demi-culverins – and fourteen smaller pieces. The demi-culverin had an extreme range, at 10° elevation, of 2,400 yards.

in pairs, geared the army to a slow pace. Among the infantry, too, those pikemen who had heavy armour were not likely to march at three miles an hour, even on a cool October day. On the other hand, the cavalry naturally thought nothing of moving twenty miles a day (after the Second Battle of Newbury King Charles and his escort actually reached Bath, fifty miles away, in less than twenty-four hours). But there was no point in the horse getting miles ahead of the foot and the rest of the army.

The converging armies blundered into each other eventually on the evening of 22 October, when quartermasters, seeking billets for the cavalry of either side, chose the same village. The Cavaliers had intended to rest on 23 October while a brigade group, as we should now call it, took the Parliamentarian-held fortress of Banbury Castle. Their plans had been well known to Essex since one Blake, who was with the King, 'betrayd all his Matyes Counsells', as the diary of Prince Rupert, the King's nephew, records. But the King and his Council of War now changed their plan and gave orders for the army to rendezvous early the next morning on the summit of Edgehill. It is unlikely that the spy had time to notify the Earl of this change.

Both armies spent the morning of 23 October (the calendar was changed in the eighteenth century and the real anniversary would today be 3 November) in assembling from their widely dispersed quarters. Essex arrayed his men in the fields between Kineton and Edgehill in full, if distant, view of his opponents. He could, no doubt, see Royalist cavalry on the summit of the ridge, but since the King's army would scarcely have paraded on the very edge of the hill, the majority must have been out of sight from the Parliamentarian position.

Essex made no move. As suggested earlier, the rebels may well have been reluctant to strike the first blow. More important, the hill on which the Royalists were drawn up was so steep as to be practically unassailable, and cavalry could not hope to ascend it in good order. Lastly, a brigade of Roundhead foot, many cannon and some troops of horse were at least a day's march behind the main army. By early afternoon the Cavaliers had assembled all their men, and, seeing that Essex would not come to them, decided to go down to him. This they did unimpeded by the Roundheads, though it was necessary to harness most of the horses *behind* the guns, so steep was the hill. The Royalists deployed in the plain, and the King, loudly cheered, rode along his line exhorting his men to do their duty. Prince Rupert gave last-minute instructions to his cavalry 'to march as close as was possible', recalls Sir Richard Bulstrode, who served in the Prince of Wales's Regiment at Edgehill, 'keeping their Ranks with Sword in Hand, to receive the Enemy's Shot, without firing either Carbin or Pistol', until their opponents were broken. They were fittingly stern orders for men about to engage in the first major battle of the Civil War.

A group of variously equipped Royalist soldiers. Armour at Edgehill ranged from that of the cuirassier (foreground, mounted), who was completely protected from head to knee, to the breast- and back-plates of the dragoon (rear left, mounted) and the pikeman; musketeers had little body protection, while some 3–400 men fought only with cudgels or pitchforks.

King Charles I
Royalist Captain-General
King Charles I (1600–49) was his own Captain-General. Just as the medieval kings of England commanded their own armies at Bosworth and elsewhere, and just as, later, King George II led his men at Dettingen, the King was the effective commander of his army at Edgehill. It is true that he had experienced generals with him, the Earl of Lindsey and Lord Forth, but the former resigned his appointment as Lord General just before the fight, and the latter did not succeed him until it was over. Edgehill was the King's first battle, but he was not entirely ignorant of military affairs, and he had been with his army ever since it was raised, taking an active part in the deliberations of the Council of War which ran its day-to-day affairs. Nobody knew better that 'money is the nerves of war', and nobody knew better how hard it was to come by. But Charles was a man of courage; he had set his course and he meant to follow it. He was not a dashing man, nor a particularly clever one, but once having made up his mind on a question involving religious or political principle he was extremely tenacious, a quality that was to stand him in good stead at Edgehill.

Earl of Essex
Parliamentarian Captain-General
Robert Devereux, third Earl of Essex (1591–1646) was not a likely military figure. He was an indifferent strategist and lethargic by nature. But he understood infantry tactics and was not afraid to risk his own skin when he deemed it necessary. It was his misfortune that he lacked the support of efficient general officers, and that his army was of a mutinous humour, little willing to stomach discipline. Essex had experience before Edgehill as a colonel soldiering in the pay of the States of Holland. His chief exploit in the English Civil War was the Gloucestershire-Newbury campaign of 1643 which can be said, without exaggeration, to have deprived the Royalists of their best chance of winning the war. In the Cornish campaign of 1644 he was less impressive, for, having led his army into a trap between Lostwithiel and Fowey, he abandoned it and escaped by sea. He died only two years later and it would be charitable to suppose that failing health had clouded his judgment in Cornwall.

The Rival Armies

The strength of the two armies was almost equal, as these brief tables show:

Royalists		**Parliamentarians**	
Cavalry	2,800	Cavalry	2,150
Dragoons	1,000	Dragoons	720
Infantry	10,500	Infantry	12,000
Total	14,300 men	*Total*	14,870 men
and 20 guns		and 30–37 guns	

At this stage of the war the Royalists were by no means completely equipped. The Earl of Clarendon in a famous passage from his *History of the Rebellion and Civil War in England* (1888) tells us that 'Amongst the horse, the officers had their full desire if they were able to procure old backs and breasts and pots, with pistols and carbines for their two or three first ranks, and swords for the rest; themselves (and some soldiers by their examples) having gotten, besides their pistols and swords, a short pole-axe.' Moreover, some three or four hundred of the Royalist foot had no better weapons than cudgels or pitchforks.

A few mounted individuals on the Royalist side – among them the Earl of Northampton, Sir Richard Willys and Captain Edward St John – seem to have worn the full cuirassier's equipment of close helmet and barred visor with heavy body-armour. On the other side the whole of Essex's Lifeguard were cuirassiers, or 'lobsters', so named after their jointed plate armour. But the normal trooper on either side was a 'harquebusier', and ideally was armed with a long, straight double-edged sword, a pair of pistols, perhaps a carbine, a buff coat, back- and breast-plates, and a lobster-tailed helmet or 'pot'.

The foot regiments of both sides comprised both pikemen and musketeers. Theoretically there were two musketeers to every pikeman, but in practice, at least in the Royalist army, there may have been roughly equal numbers of each type of foot-soldier.

Some of the troops, equipped by wealthy commanders, must have looked very fine. The King's Lifeguard was nicknamed 'The Troop of Show'. Another, thought from 'the bravery of their accoutrements' to have been raised by the Earl of Newcastle, in which case it was part of the Prince of Wales's Regiment, is described by a contemporary observer as riding 'fifty great horses all of a darke Bay, handsomely set out with ash-colour'd ribbons, every man gentilely accoutred, and armed'.

There was a certain amount of uniformity within regiments since some articles, including caps, coats, breeches and stockings were general issues. The cavalry of both sides normally wore buff coats and scarves, rose-red for the Cavaliers and orange-tawny, the colours of the Earl of Essex, for the Roundheads.

The Earl of Essex arrayed his men in the fields between Kineton and Edgehill, while the Royalists formed up in a virtually unassailable position on the summit of Edgehill. The Royalists later went down to the enemy. In the model view (below) the rival armies are seen from behind the Parliamentarian position, looking uphill towards Radway.

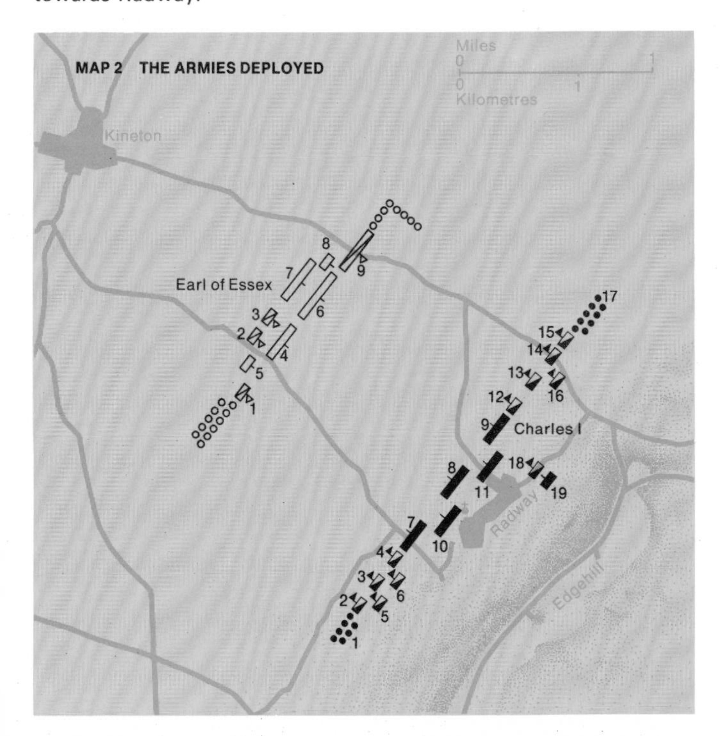

ROYALISTS

Infantry

Cavalry

Dragoons

1 Aston
2–6 Wilmot's cavalry wing
2 Wilmot's Regt
3 Grandison
4 Carnarvon
5 Digby
6 Aston's Regt
7–11 Astley's infantry brigades
7 Wentworth
8 Richard Feilding
9 Gerard
10 Nicholas Byron
11 Belasyse
12-16 Rupert's cavalry wing
12 Maurice
13 Rupert's Regt
14 Prince of Wales
15 King's Lifeguard
16 John Byron
17 Usher
18 Gentlemen Pensioners
19 Legge's Firelocks

PARLIAMENTARIANS

Infantry

Cavalry

Dragoons or musketeeers

1-3 Balfour's cavalry wing
1 Lord Feilding
2 Stapleton
3 Balfour's Regt
4 Meldrum's brigade
5 Fairfax's Regt (part of Meldrum)
6 Charles Essex
7 Ballard
8 Holles's Regt (part of Ballard)
9 Ramsey's cavalry wing

There were several instances during the war of officers being taken prisoner through mistaking an enemy regiment for one of their own. Since each side adopted the same system for the colours of their infantry companies this is scarcely surprising. The wearing of field-signs, sprigs of oak, white bands or pieces of paper in their hats, was a common device at this period, intended to prevent mistakes of identification.

Sir John Heydon commanded the Royalist artillery. This consisted of twenty pieces, but of those only six were big guns (two 27-pounder demi-cannons, two 15-pounder culverins and two 9-pounder demi-culverins, the last having an extreme range of 2,400 yards). These evidently formed a single battery. The other fourteen were doubtless employed well forward, in close support of the infantry brigades.

Details of the Parliamentarian artillery, which was commanded by the Earl of Peterborough, are few, but Essex seems to have had at least thirty guns and possibly as many as thirty-seven. Some were sakers; this was a useful 5¼-pounder field-gun with a point-blank range of 360 yards and an extreme range of 2,170.

In practice, fire was seldom opened at more than 1,000 yards owing to the difficulty of observing the effect at a time when few officers had 'perspective glasses' and powder-smoke often obscured the target.

Organization

The Royalists had nine horse-regiments as well as two troops that were not regimented (belonging to the King's Lifeguard and the Gentlemen Pensioners). The biggest regiment, the Prince of Wales's, with eight troops, probably numbered 500, whilst those of Lord Digby and Sir Thomas Aston probably did not exceed 150 apiece. The cavalry was not yet brigaded but was simply divided into two wings, the right under Prince Rupert and the left under Lord Wilmot, the former being considerably stronger. Estimated strengths are 1,695 for Prince Rupert and 1,055 for Lord Wilmot. The Royalists also had three regiments of dragoons.

The Royalist foot-soldiers, under the command of Sir Jacob Astley, were in five brigades or tertias. Their regiments varied in strength, some having as many as ten companies, others as few as seven.

Brigade Commander	Number of Regiments	Estimated strength on 16 November 1642[1]
Charles Gerard	3	1,985
John Belasyse	3	1,880
Richard Feilding	5	2,545
Sir Nicholas Byron	3[2]	1,830
Henry Wentworth	3	1,790

[1] Calculated from a pay warrant. Numbers at Edgehill were probably rather greater.

[2] Including the King's Lifeguard of Foot.

The Parliamentarians do not seem to have completed the process of regimenting their numerous troops of horse, but elements of at least eight regiments can be identified – although one of these, that belonging to Lord Willoughby of Parham, was with Colonel John Hampden's brigade, a day's march behind the army.

There were three regiments, each of six troops as it happened, on the Roundhead right – those of Lord Feilding, Sir Philip Stapleton and Sir William Balfour. On the left wing there were twenty-four troops, and it is attractive to suppose that they were organized in four regiments, but this is probably too tidy a solution. Sir William Balfour commanded the horse of the right wing and Sir James Ramsey that of the left.

The Roundheads placed both their regiments of dragoons, a total of 720 men, on the right wing, and used musketeers drawn out of the foot-regiments to line hedges on their left, presumably because they had too few dragoons to protect both flanks. The dragoon colonels were both Scots: John Browne, who had been defeated at Powick Bridge, and James Wardlawe.

The Parliamentarians had twenty regiments of foot but only twelve were in the battle. Two more reached Kineton before nightfall. Essex had frittered away his strength by putting regiments into garrison at Banbury Castle, Coventry, Worcester, Hereford and, possibly, Northampton. Moreover, Hampden's brigade was away escorting artillery somewhere between Worcester and Kineton. This left Essex three brigades, each of four regiments. The brigade commanders were Sir John Meldrum, Colonel Charles Essex and Colonel Thomas Ballard.

The regiments in theory were 1,310 strong, including officers, and had they been up to strength Essex would have had 15,720 foot in the field, with another 2,620 due before nightfall. But there had been a certain amount of sickness and desertion, and one regiment, Ballard's, ordered into the field before it was complete, had only 800 men. It is evident, however, that this was exceptional and an average of 1,000 men per regiment seems a moderate computation for the Roundhead foot. If this is correct the brigades were about 4,000 strong, as opposed to those of the King, which with one exception were manned by 2,000, or even slightly less.

The Course of the Battle

The first phase of the battle was an ineffective artillery duel, which did little damage to either side. Then, at a given signal, probably the simultaneous discharge of all their heavy ordnance, the Royalists advanced. Their dragoons cleared both flanks, disposing of the Parliamentarian dragoons and musketeers opposed to them. Rupert charged Ramsey, whose men opened fire at long range and then turned rein. Wilmot quickly swept Feilding aside, but missed Balfour and Stapleton, presumably because they were masked behind Meldrum's brigade. On his wing

Prince Rupert's cavalry wing sweeps past the Parliamentarian left. By failing at that point to swing into the enemy's vulnerable flank, the Royalists missed a great opportunity of there and then destroying the first Roundhead army. Remnants of Sir James Ramsey's defeated cavalry are shown in flight.

Rupert managed to stop three troops and on Wilmot's Sir Charles Lucas rallied some 200 men, but of the Royalist horse six men out of every seven, including the second line of each wing, were now happily pursuing fugitives and plundering baggage-wagons in the streets of Kineton. Lord Digby was but a young soldier, but Sir John Byron should have known better. Had he been able to swing his regiment into the enemy's left flank the Parliament's first army might have been destroyed that day. It is not so far-fetched a theory, for, when Ramsey fled, Charles Essex's brigade with few exceptions also departed, leaving its commander to his fate. Altogether the Parliamentarians now had left two regiments of horse and seven regiments of foot (one of which, Holles's, was bravely rallying, having been ridden over by its own horse during Ramsey's retreat). Fortunately for the Parliamentarians the commander of the remaining horse, some 650 strong, was Sir William Balfour, a bold and experienced general.

By this time Astley had marched up with his five Royalist tertias and they were at 'push of pike' with Meldrum and Ballard. The fight was long and indecisive. As Clarendon tells us: 'The foot of both sides stood their ground with great courage; and though many of the King's soldiers were unarmed and had only cudgels, they kept their ranks, and took up the arms their slaughtered neighbours left them; and the execution was great on both sides, but much greater on the earl of Essex's party. . . .' Some 10,500 Royalist foot were struggling to dispose of about 6,400 Roundheads – those who had not fled. Perhaps in time numbers would have told, but both sides were somewhat unpredictable, being composed of young and amateur soldiers.

James II (1633–1701), younger brother of the Prince of Wales, was also present, and questioned many of the survivors; later he wrote: 'The foot being thus engaged in such warm and close service, it were reasonable to imagine that one side should run and be disorder'd; but it happened otherwise, for each as if by mutual consent retired some few paces, and they stuck down their colours, continuing to fire at one another even till night; a thing so very extraordinary, that nothing less than so many witnesses as there were present could make it credible.' James's account is of interest and is doubtless completely accurate in so far as the brigades of Gerard, Belasyse and Wentworth were concerned. But a ruder experience lay in store for Feilding's men. They were attacked and broken by Balfour, who came out through gaps in the Parliamentarian line, and charged right into the main Royalist battery. The Prince of Wales, then aged twelve, who was with the Gentlemen Pensioners, narrowly escaped capture ('I fear them not!' he cried as he spanned his wheel-lock pistol). Balfour failed to spike the captured Royalist guns for lack of nails. Withdrawing his men he made a concerted attack of horse and foot upon Sir Nicholas Byron's brigade, which, after a stiff resistance, was broken. Lindsey fell, his son Lord Willoughby d'Eresby, Colonel of the Lifeguard of Foot,

left Prince Rupert and his faithful dog, who accompanied him in battle. **below** Astley's five Royalist infantry brigades closing with those of Meldrum and Ballard. The fight was long and indecisive and lasted until nightfall. Clarendon wrote that 'The foot of both sides stood their ground with great courage; and though many of the King's soldiers were unarmed and had only cudgels, they kept their ranks, and took up the arms their slaughtered neighbours left them' (*History of the Rebellion and Civil War in England,* 1888).

was taken while hastening to his assistance, and the Banner Royal was snatched from the lifeless hand of Sir Edmund Verney, the Knight Marshal.

The battle had now taken a decisive turn in the Roundheads' favour. Indeed for a time it looked as if they might have snatched victory from defeat, for there was a great gap in the Royalist line where the brigades of Feilding[1] and Byron had been.

Even so, Gerard's, Belasyse's and Wentworth's brigades must still have been at least as strong as those of Meldrum and Ballard, which must have been just as exhausted as they. At this crisis the King himself rode down and inspired his men to stand fast. It seems not unlikely that Wentworth moved his brigade to his right in order to close with Belasyse's left flank.

Meanwhile Sir Charles Lucas had led his 200 horse into the right rear of the Roundhead array. The charge spent itself among the runaways of Charles Essex's brigade and the regiment of Sir William Fairfax, but it was not fruitless, for it ended with the recapture of the Banner Royal. To retake it Captain John Smith attacked six horsemen who were escorting the captured banner. He was wounded, but he killed one Roundhead and wounded another. The rest fled. He also rescued Colonel Richard Feilding, and was later knighted for his services.

Perhaps the turmoil of Lucas's charge made some of Meldrum's and Ballard's men look over their shoulders. They did nevertheless advance some distance after Byron's brigade was broken. They were stopped by the return of a number of the Royalist horse, albeit in disorder; by case-shot from the big guns, and by the fact that the Royalist foot had now 'made up a Kind of a Body again', in the words of what may be called the Parliamentarian Official Account. The Parliamentarian left was repulsed by dragoons, presumably Usher's. Lord Falkland even urged Wilmot, who had now returned, to charge Balfour, but Wilmot answered: 'My Lord, we have got the day, and let us live to enjoy the fruit thereof.'

It was getting dark, and both sides had reached the end of their tether. The Royalists, as their Official Account states, 'durst not Charge for fear of mistaking Friends for Foes (though it was certainly concluded . . . that if we had had light enough, but to have given one Charge more, we had totally routed their Army).'

The Parliamentarian version is equally disingenuous: 'And by this time it grew so dark, and our Powder and Bullet so spent, that it was not held fit that we should Advance upon them; but there we stood in very good order; drew up all our Forces . . . and so stood all that night upon the place where the Enemy, before the Fight, had drawn into Battalia. . . .'

The truth is that both sides had had enough. There

[1] There is a vague clue that Colonel Sir Edward Fitton's regiment, of Feilding's brigade, was still in action protecting the artillery.

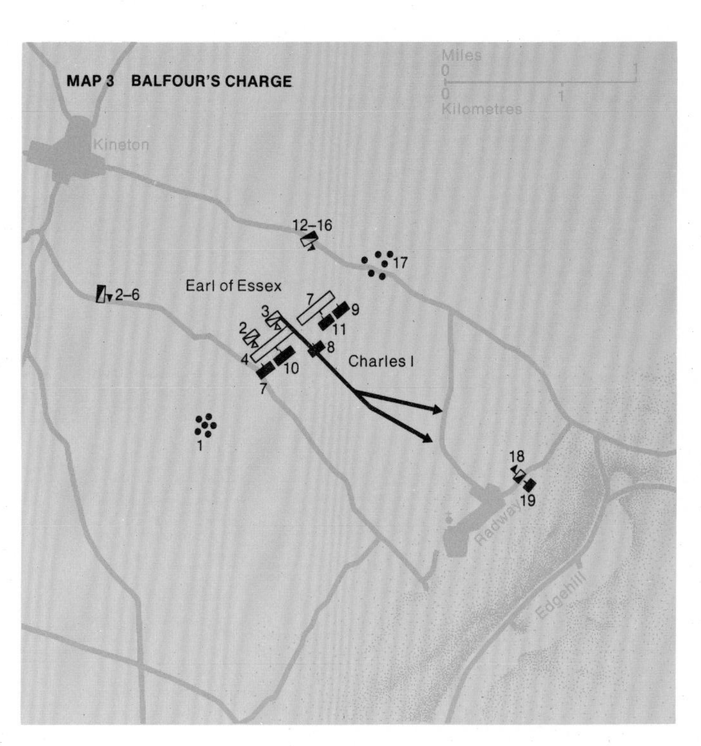

The greater part of the Royalist cavalry, having swept Lord Feilding's Regt and Ramsey's wing from the field in their opening charge, rode wildly on to Kineton, pursuing fugitives and plundering baggage-wagons. Of the Roundhead infantry Fairfax's Regt and most of Charles Essex's brigade had fled, but the rest stood firm against Astley's Royalist brigades. Then Sir William Balfour led a bold cavalry charge through the Royalist centre, overrunning first Richard Feilding's brigade and later Sir Nicholas Byron's. Balfour's initiative turned the battle strongly in the Parliamentarian's favour.

ROYALISTS

Infantry

Cavalry

Dragoons

1 Aston
2-6, 12-16 Units of Royalist cavalry under Lucas and Rupert respectively
7 Wentworth
8 Richard Feilding, overrun by Balfour's men
9 Gerard
10 Nicholas Byron
11 Belasyse
17 Usher
18 Gentlemen Pensioners
19 Legge's Firelocks

PARLIAMENTARIANS

Infantry

Cavalry

2 Stapleton
3 Balfour's Regt
4 Meldrum
7 Ballard

followed a night 'as cold as a very great frost and a sharp northerly wind could make it'.

In the evening, Colonel John Hampden had arrived with two regiments of foot and Lord Willoughby of Parham's regiment of horse, besides other troops, including Captain Oliver Cromwell's, as well as guns. But the Earl, whose army had been severely crippled by the unceremonious departure of thirty troops of horse, two regiments of dragoons and five of foot, was not eager to renew the encounter. The two sides, suffering from hunger and cold, gazed upon each other for three or four hours the following day, then both departed, Essex to Warwick and the Cavaliers to the quarters they had occupied before the battle. On the morning of the 25th Rupert fell on the Roundheads' rear and took twenty-five wagons; four powder-carts were blown up. A consequence of this attack was that Essex lost his plate and his cabinet of letters, which revealed that the spy, Blake, had been betraying the King's secrets for £50 a week. Blake was hanged.

By withdrawing, Essex left the battlefield to the King, with seven cannon (two 12-pounders, one 6-pounder and four 3-pounders) making a respectable addition to the Royalist train of artillery.

In the battle the Parliamentarians had taken sixteen Royalist foot colours. The Cavaliers had captured at least fifty-two cornets and colours. 'I believe ther be manie

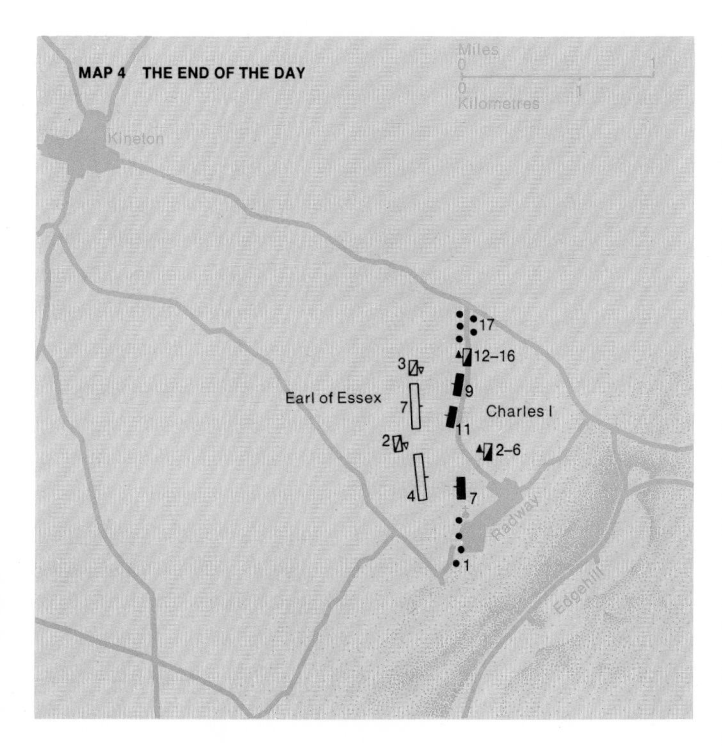

The fighting ended when both sides, cold, hungry and exhausted, decided that there was little to be gained from continuing in the fast-vanishing light. Losses totalled about 1,500 killed on both sides.

ROYALISTS

Infantry

Cavalry

Dragoons

1 Aston
2-6 Part of Wilmots' cavalry wing
7 Wentworth
9 Gerard
11 Belasyse
12-16 Part of Rupert's cavalry wing
17 Usher

PARLIAMENTARIANS

Infantry

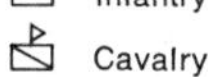
Cavalry

2 Stapleton
3 Balfour's Regt
4 Meldrum
7 Ballard

more,' wrote Sir Edward Sydenham, the new Knight Marshal.

Losses in some of the Roundhead regiments had been heavy, especially in those that ran. Sir Henry Cholmley had 1,200 men on 1 October and in late November, no more than 552, though some may have deserted. Ballard had 776 on 17 October but only 439 on 11 November. It is thought that about 1,500 of both sides were killed. The Roundhead casualties included at least one woman, as the Parish Register of Little Brickhill, on Essex's route from Warwick to London, records: 'Agnes Polter, wounded in the battel of Edgehill was buryed.' Perhaps she was an unlucky camp-follower cut down when the Cavaliers charged through the streets of Kineton and plundered the baggage-train.

Aftermath and Conclusion

The Royalists took some time to organize themselves after the battle. Key officers had fallen, there were wounded to be cared for, arms to be collected, and all that multitude of administrative chores which take time even in an experienced army. On 27–8 October the Cavaliers took Banbury and Broughton Castle. On the 29th they entered Oxford, carrying the captured colours in triumph. Oxford was a valuable prize, and it remained the King's capital for nearly four years.

On 4 November the Royalists entered Reading and on the 7th Rupert called upon Windsor Castle to surrender and was refused. On the same day Essex arrived back in London, having marched there via Dunstable and St Albans, and the King's chance of capturing the City eluded him. On 12 November the Cavaliers stormed Brentford, destroying the regiments of Holles and Lord Brooke, but next day they were faced on Turnham Green by Essex's Army and the London Trained Bands, a total of 24,000 men. The Royalists, seriously outnumbered, declined battle. On 19 November the King withdrew to Reading and early in December sent his army into winter quarters.

The Royalists, lacking the financial resources of London, could only hope to win if they could make the war a short one. They needed a decisive victory. Edgehill was a victory, but only by a relatively narrow margin. And though it gave the King Oxford, an asset in the long war he did not want, it did not give him London.

left Rival cavalrymen contest a standard. At one point, following Balfour's charge, the Banner Royal was seized in a Roundhead attack. It was recaptured after a brief interval by Captain John Smith, who drove off six men escorting the banner, killing one; he was knighted for this and other services at Edgehill.

David Chandler at

BLENHEIM

In this crucial conflict in the War of the Spanish Succession the Duke of Marlborough struck up a military partnership with Prince Eugène of Savoy that proved irresistible to the numerically greater Franco-Bavarians under Marshal Tallard. Marlborough's action in this battle, while establishing him throughout Europe as a commander of genius, saved the Second Grand Alliance from collapse.

1704

As dusk gathers, Lord Orkney's red-coated infantry battalions close in for the last time on Blenheim village. Following a parley the French infantry, deserted by their leader, Clérambault, surrendered.

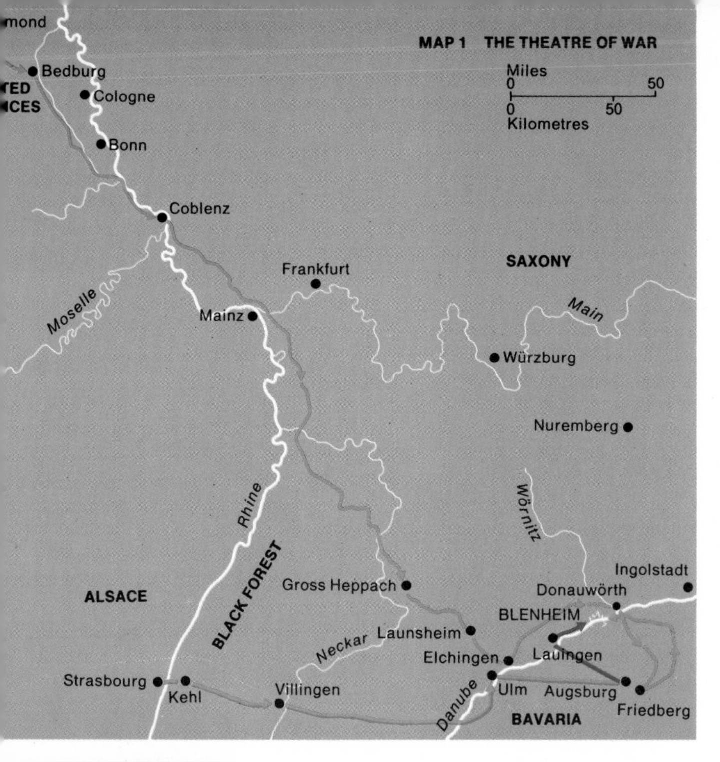

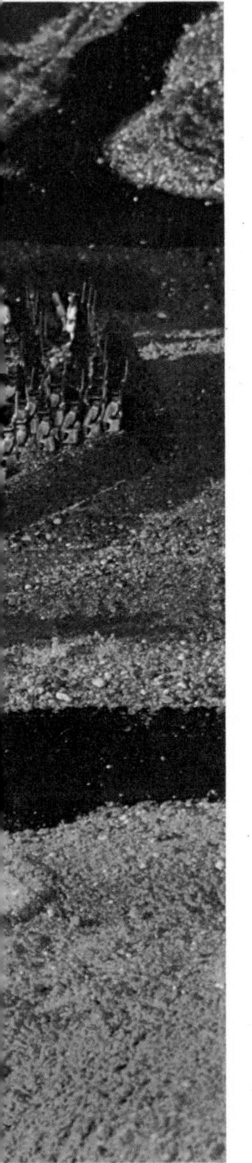

The early years of the War of the Spanish Succession (1701–14) were fought on three main fronts: in northern Italy, the Netherlands, and in a third zone stretching from the Middle Rhine to the Upper Danube. The Battle of Blenheim sprang from the Duke of Marlborough's resolve to help his threatened ally, Leopold I of Austria. To do this he transferred, in five weeks, an army that eventually totalled some 40,000 men over a distance of 250 miles from the Netherlands to link with Austrian forces between Gross Heppach and Launsheim. He then took Donauwörth and invaded Bavaria, returning ahead of the advancing Franco-Bavarian army to join up with Prince Eugène of Savoy near Münster, about five miles from Blenheim (modern Blindheim).

Main Allied forces (Marlborough)
French (Tallard)
Franco-Bavarians (Tallard/Elector)

The Background to the Battle

From 1702, England was deeply involved in the War of the Spanish Succession. As an original signatory – with the United Provinces of the Netherlands and Austria – of the Second Grand Alliance Treaty, the government of Queen Anne was committed to an eleven-year struggle against the Bourbons – Louis XIV of France and his grandson, Philip, Duke of Anjou, who had recently been bequeathed the entire Spanish inheritance after years of intrigue.

For half a century France had been the bully of Europe, and if any semblance of an equitable balance of power was to be maintained it was widely felt outside France that a fair partition of the Spanish possessions (which included the area of modern Belgium and much of northern Italy besides the Philippines and vast possessions in the Americas) was a matter of international importance. In addition to the principal Allies already mentioned, most of the myriad states of the Holy Roman Empire followed Austria's lead, save only Bavaria and Cologne, which espoused the Bourbon cause in September 1702. By the end of the following year, both Savoy and Portugal had joined the Grand Alliance, and the rival camps assumed their final proportions.

During the first years of the war, fighting centred on three main theatres. In northern Italy, Prince Eugène of Savoy, an Imperial Austrian general, was hard-pressed to withstand the superior forces of Marshals Catinat and Vendôme, but eventually managed to hold his ground. In the Netherlands, John Churchill, created Duke of Marlborough in 1702, fought two generally successful, if conventional, campaigns along the Rhine (1702) and the Meuse (1703), establishing a secure southern flank for the United Provinces. On the third front, which stretched from the Middle Rhine to the Upper Danube, the French lost no time in reinforcing their new ally, Elector Max Emmanuel of Bavaria, and by dint of hard fighting succeeded in establishing control over most of the immediate region by late 1703.

Here, as Marlborough and a few other enlightened statesmen realized, lay the maximum peril. The Emperor of Austria, Leopold I, his capital already threatened by Hungarian patriots in full revolt to the east, now faced a new double threat – from the Elector and the French operating along the Danube valley, and from Vendôme striking northwards into the Tyrol over the Brenner Pass. The fall of Vienna would mean the collapse of the Alliance, and the Emperor urgently entreated aid.

The call did not go unheeded. Despite immense political, military and logistical difficulties, Marlborough set out in May 1704 to lead part of his army from the Netherlands to the Danube, planning to forestall his opponents and drive Bavaria out of the war. 'I am very sensible that I take a great deal upon me,' he wrote, 'but should I act otherwise the Empire would be undone, and consequently the Confederacy.' By an extraordinary combination of guile and skill, he hoodwinked or calmed his over-cautious Dutch allies, fooled a watchful but over-confident foe, and in the brief span of five weeks transferred an army that eventually totalled 40,000 men of many nations over a distance of 250 miles to link up with the Imperial commanders – Prince Lewis of Baden and Prince Eugène (brought in at Marlborough's specific request) – between Gross Heppach and Launsheim in south-western Germany. His task was completed by 21 June.

Next, the Duke despatched Prince Eugène with a skeleton force to watch the Rhine where more French reserves were massing under Marshal Tallard prior to reinforcing Bavaria. He himself then led Prince Lewis of Baden in a bold dash to the Danube, intent on defeating the Elector of Bavaria in one blow. The result was the storming of the Schellenberg Heights outside Donauwörth on 2 July, which gave the Allies a bridge over the Danube into Bavaria and also a terminal for Marlborough's new system of communications. This ran back through friendly territory to Nördlingen and the River Main.

An ensign and a fusilier in the service of France in the early eighteenth century. By 1704 most European armies were, broadly speaking, similarly equipped. The 'poor foot' carried muzzle-loading flintlock muskets fitted with socket bayonets (the last pikes having been replaced as recently as 1703). The English version, 'King William's land musket', was of .85-inch calibre, fired a one-ounce ball and could be reloaded twice a minute. The French *fusil* was of .68-inch calibre and fired twenty-four balls to the pound. Accurate ranges were not much more than a hundred yards, and most fire-fights were engaged at sixty yards.

Success now eluded the Duke. The Elector refused either to fight or to negotiate, but withdrew inside his fortresses, which the Allies, without heavy artillery, could not besiege. Aware that Tallard was on the point of marching through the Black Forest to his aid, Max Emmanuel preferred to sit tight whilst the Allies ravaged his countryside, burning possibly 400 villages. By mid-July Eugène was falling back with only 9,000 men from the Rhine area, shadowing the ponderous advance of Tallard's French army, and on 5 August the Marshal at last reached Augsburg. The strategic initiative had now indubitably passed to the French and the Bavarians. While Marlborough hastily conferred with his colleagues, the combined Franco-Bavarian army proceeded to recross the Danube at Lauingen, and then conducted a leisurely advance eastwards towards Donauwörth. Were the bridges there to fall into Tallard's hands, Marlborough would effectively be cut off south of the Danube, with his communications to northern Germany and the United Provinces severed.

Marlborough's quandary was acute. The French had almost sprung a trap around him; even so, Baden was disinclined to risk a battle, although both the Duke and Prince Eugène realized that nothing but the boldest gamble could retrieve the situation. On 7 August the decision was taken. Marlborough detached Baden with 15,000 men to besiege Ingolstadt – the next bridge below Donauwörth – a course of action in which Baden grumblingly acquiesced, whilst Marlborough, by dint of forced marches, rushed the remainder of his army northwards out of Bavaria. By 11 August the transfer was complete, and the Duke's men had joined Eugène's slender force near Münster, making a joint total of 52,000 men and sixty-six guns. On this date, the French moved into a comfortable camp some five miles

French horse and foot soldiers of the period. The colours of uniforms were gradually becoming standardized by this time. The English wore red or blue (ordnance personnel), the Dutch dark grey or brown, the Austrians pearl grey, the French white, red or blue, and the Bavarians sky-blue. Regimental facings were already customary.

The Duke of Marlborough
Allied Commander-in-chief
John Churchill, First Duke of Marlborough (1650–1722) was the son of a Devonshire squire. In 1665 he became a page to the Duke of York, and soon after was commissioned into the Foot Guards. Twenty-five years of varied military and political experience followed. In 1674 he married the fiery Sarah Jenyns, favourite of the Princess Anne, and established an important personal and political relationship with the latter. Under William and Mary (1689–1702) he was made Earl of Marlborough, and in 1691 in southern Ireland he commanded his first independent force with great success. On Queen Anne's accession in 1702, he was confirmed as Captain-General of the forces in Holland, and proceeded to demonstrate his full military abilities. By 1704 he was a widely-respected military figure, though not an internationally famous one.

In character he was bland, courteous and polished; he was also highly ambitious and a strong avaricious streak ran through his nature, although he could act unselfishly when there was need. His charisma affected all his troops, whether native or foreign. A strategist and tactician of genius, his skill at administration and his genuine concern for his men's welfare earned him the nickname of 'the Old Corporal'.

Marshal Tallard
Franco-Bavarian Commander-in-chief
Marshal Camille d'Hostun, Marquis de la Baume and Comte de Tallard (1652–1728), was a French commander of distinguished ancestry and fair ability. He was also an experienced diplomat: whilst serving as French ambassador to the Court of St James in London he negotiated both Partition Treaties (1698 and 1700) with William III.

After his victory in the battle of Spire (or Speyerbach) near the Rhine in 1703, he was rewarded with his marshal's baton by Louis XIV, and the following year he was put in command of an army with orders to reinforce the Elector of Bavaria, and mount an advance down the Danube towards Vienna. The campaign of 1704 would result in his complete eclipse as a commander, but after seven years as a captive in England he was repatriated, and after the accession of Louis XV became a member of the Council of Regency (from 1717) and ultimately a Minister of State (1726).

In contrast to Marlborough, he was renowned for his hospitality to his officers, even on campaigns; but he was proud and pompous and prone to histrionics and emotional outbursts. As a commander he tended to be lethargic and ultimately proved incapable of controlling his subordinate generals.

away near the village of Blenheim (today Blindheim), never suspecting for an instant that within forty-eight hours they would be locked in a life-or-death struggle.

That, however, was what the Duke and the Prince were determined to bring about. After conducting a careful reconnaissance of the French camp from the spire of Tapfheim church on 12 August, they issued their orders. Despite their numerical inferiority (Tallard disposed of 56,000 men and all of ninety cannon), and the obvious strengths of the enemy position, the 'Twin Princes' of the Grand Alliance were determined to venture all on a major battle the next day. They were playing for high stakes, but each had an implicit trust in the other, and both knew the calibre of the troops under their command.

So it was at 3 a.m. on Whit Sunday, 13 August 1704, as an unsuspecting Tallard and his army lay snug in their tents and billets, that nine Allied columns issued silently forth from Münster and Tapfheim and began to close on the enemy camp behind the River Nebel. Their movement was shrouded in a dense early-morning mist; Marlborough led the left and centre of the Allied force; Eugène and the right wing set off on a wider arc through hilly and wooded country to form the northern extremity of the Allied line. One of the most notable battles of European history was about to be launched.

The Rival Armies

The comparative strengths of the opposing armies on 13 August 1704 were approximately as follows:

Franco-Bavarians		**Allies**	
Cavalry	14,500	Cavalry	17,800
Infantry	39,000	Infantry	33,000
Train personnel (artillery-men and engineers)	2,500	Train personnel (artillery-men and engineers)	1,200
Total		*Total*	
56,000 men and 90 guns		52,000 men and 66 guns	

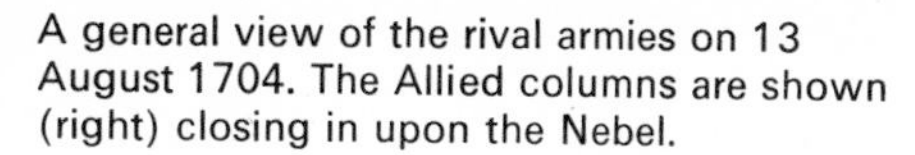
A general view of the rival armies on 13 August 1704. The Allied columns are shown (right) closing in upon the Nebel.

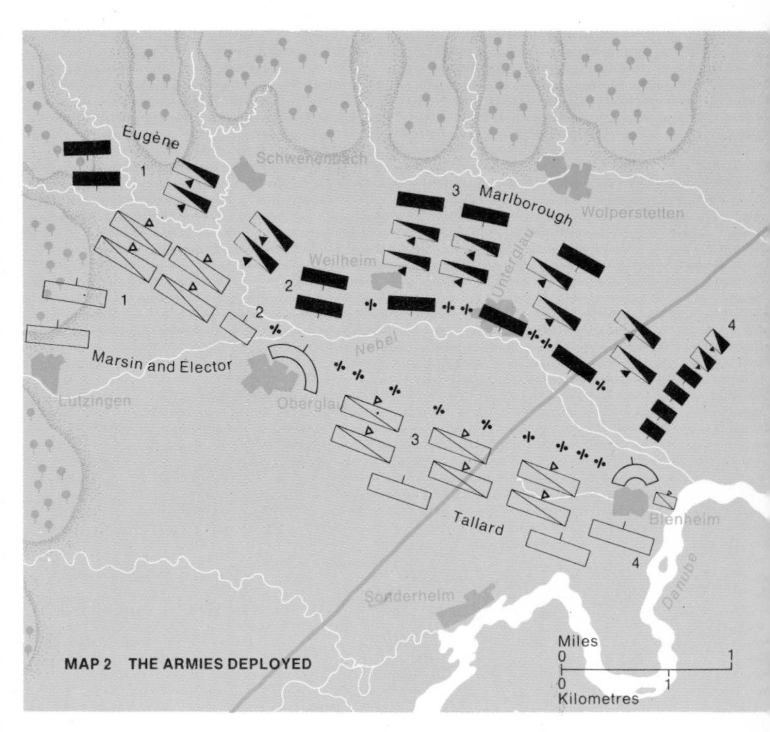

Tallard arrayed his army in a strong position, its right flank protected by the Danube, its left by a series of wooded ridges. To the front ran the Nebel, a tributary of the Danube, across which Tallard hoped to lure Marlborough before attacking him in the flanks from Blenheim and Oberglau. The Allied army was positioned to attack Blenheim on the left, while part of the centre was directed to assault Oberglau, the remainder to cross the Nebel and aim for Tallard's centre. The right wing, under Prince Eugène, placed against Marsin and the Elector, arrived some four hours behind the main Allied body after a longer and more difficult march. Battle commenced at 12.30 p.m.

ALLIES

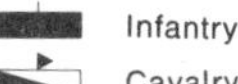
Infantry
Cavalry
Main batteries

1 Eugène (18 infantry battalions, 92 cavalry squadrons)
2 Holstein-Beck (10 centre bns under Charles Churchill)
3 Charles Churchill's main body (18 bns, 72 sqns)
4 Cutts (20 bns, 14 sqns)

FRANCO-BAVARIANS

Infantry
Cavalry
Main batteries

1 Marsin and Elector's main body (28 bns, 67 sqns)
2 De Blainville (14 bns)
3 Zurlauben (9 bns, 64 sqns)
4 Clérambault (9 bns in Blenheim, 18 bns and 12 sqns to rear and right of village)

The Franco-Bavarian force, of which about a quarter was Bavarian, was divided into seventy-eight infantry battalions and 143 cavalry squadrons. At this period a French battalion comprised on paper some 690 officers and men; two or more went to make up a regiment, and each battalion was divided into twelve 'line' and one 'grenadier' companies. Campaign strength was often nearer 500 men than 700. A French cavalry regiment comprised some twelve companies of three or four officers and thirty-five troopers apiece; for battle, *ad hoc* squadrons were formed by grouping three or four companies together; a squadron might contain 100–150 horsemen all told. At Blenheim, however, a considerable number fought on foot because of an epidemic which had killed many of the horses. As for Tallard's artillery, most of his pieces were 16-, 8- and 4-pounders, but there were also four 24-pounders.

Tallard certainly did not expect a serious battle to develop on the 13th – as late as 7 a.m. he added a postscript to the previous day's report to Versailles indicating that although he knew the Allies to be on the move (indeed his

The Duke of Marlborough, summoned to the scene of the crisis in the Allied centre, leads up reserves to assist in the struggle for Oberglau.

outposts had been in action since 6 a.m.) he expected them 'to march on Nördlingen this day'. By nine o'clock he had realized his error and ordered the general alarm to be sounded.

Tallard's order of battle was evidently designed to make the most of the natural advantages of his strong position. The right flank was protected by the Danube – crossed by a single pontoon bridge – and the extensive marshes surrounding it; the left by a series of wooded ridges. To the front ran a small tributary of the Danube, the Nebel, which had marshy banks and flowed into the Danube almost at right angles. Three villages – all prepared for defence – overlooked the stream at intervals along the four-mile front, namely Blenheim on the right, Oberglau in the centre, and Lutzingen a little to the rear of the left. Some 800 yards beyond the Nebel there rose a low but dominating ridge, linked by a series of water-meadows to the hamlet of Sonderheim on the banks of the Danube.

The alarm posts taken up by Tallard's army reveal his probable plan of battle. On the extreme right, between Blenheim and the Danube, he posted twelve squadrons of dismounted dragoons. In Blenheim itself he posted nine battalions, and to the rear of the village were drawn up the infantry reserves in two bodies of seven and eleven battalions respectively. Lieutenant-General the Marquis de Clérambault was sector commander. In the centre, sixty-four squadrons of cavalry were massed in two lines under General Zurlauben, supported by nine newly-raised battalions and several batteries. Apart from sixteen squadrons drawn from Marshal Marsin, these troops of the right and centre comprised the forces Tallard had marched from Alsace. From the village of Oberglau to the extreme left were deployed the main forces of Marsin and the Elector of Bavaria. Fourteen battalions under General de Blainville garrisoned Oberglau; to its west were drawn up sixty-seven squadrons and a dozen battalions, and holding the extreme flank were sixteen more battalions of white-coated French and blue-coated Bavarians. The artillery of the two armies, French and Bavarian, was sited along the battle-line, with heavy concentrations near Blenheim and Oberglau.

Tallard's plan – now that he was committed to fight – may be outlined in this way. He hoped to lure Marlborough to cross the Nebel in the centre, and then, while the Allies struggled to regroup in the difficult marshy conditions, to attack them in the flanks, from Blenheim and Oberglau. Thereafter, a downhill attack by the massed French squadrons could be hoped to fling the Allies back into the Nebel in complete devastation. If the remaining Franco-Bavarian sectors meanwhile held firm, the battle would then have been largely won. This was a sound plan – had it not been compromised by certain important errors of judgment. Firstly, the deployment of the two forces in Blenheim and Oberglau was a bad error; not only did this create two virtually self-contained units, but co-operation between left and right would anyway be hard to achieve, especially given the acrimonious relations existing between Tallard and his fellow-commanders (relations largely caused by Tallard's own overbearing attitude).

Secondly, this plan deliberately surrendered the initiative from the very outset, since its success depended on Marlborough being lured into making the first move. But Marlborough did not immediately respond. Instead almost four invaluable hours (to Tallard) were wasted through inaction, during which time the Allies were able to complete the deployment of their army, which remained until well after midday bereft of its right wing.

The Allied army, 52,000 strong, comprised sixty-six battalions, 178 squadrons and sixty-six cannon. The English component numbered fourteen battalions and seven mounted regiments, or approximately one-fifth of the whole. The redcoat battalions had a paper establishment of 780 rank and file, besides officers and regimental staff. The Dutch, Danes and Austrians tended to adopt the multi-battalion regiment, some of which held well over 1,000 men. The cavalry regiments contained some twelve troops of about fifty officers and men apiece, but reorganized for battle into squadrons; dragoon formations were often larger. However, all cited strengths need to be treated with caution; for battle strength was often one-third less than paper strength.

Marlborough, aware of the possible pitfalls of the situation, had devised an unusual order of battle for the occasion. It would be over-bold to attribute to him a full, predetermined plan of action, but he must have had certain broad ideas in mind after his reconnaissance and from information brought in by deserters. On his left, he placed twenty battalions (including the majority of the English foot) and fourteen squadrons, all under Lord Cutts, with orders to test, and if practicable to rush, the defences of Blenheim village. In the centre he entrusted his brother, General Charles Churchill, with a four-line-deep array. In front were seventeen battalions (including ten under the Prince of Holstein-Beck); behind were drawn up two lines of cavalry, totalling seventy-two squadrons; and in rear stood a further eleven battalions. Part of this force was

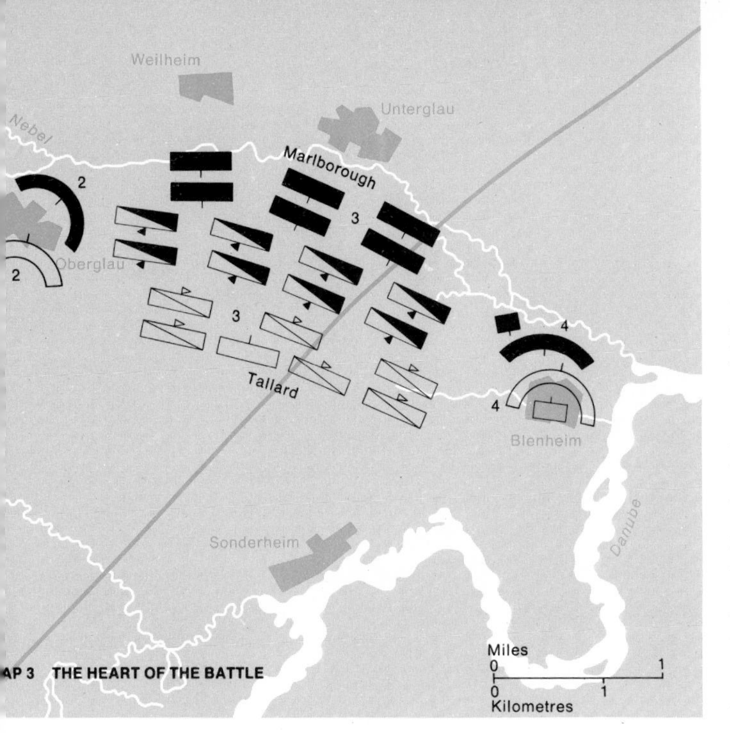

By 4 p.m., in the centre of the battlefield, a total of eighty Allied squadrons and twenty-two battalions was facing sixty and nine respectively of the French, whose garrisons at Oberglau and Blenheim had been effectively sealed. The moment of decision had almost arrived.

ALLIES

Infantry

Cavalry

2 Holstein-Beck's infantry (10 battalions)
3 Charles Churchill's main body (22 bns, 80 sqns)
4 Cutts (16 + 4 bns)

FRANCO-BAVARIANS

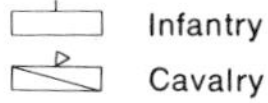

Infantry

Cavalry

2 De Blainville (14 bns)
3 Zurlauben (9 bns, 60 sqns)
4 Clérambault (27 bns)

deputed to assault Oberglau; the rest, horse, foot and guns, were to cross the Nebel and attack Tallard's centre.

Both the Allied left and centre were drawn up by 9.30 a.m., but Eugène and the right wing could not come into line before midday, owing to the greater distance they had to travel and the difficult terrain they had to cross. In due course, however, the Prince deployed ninety-two squadrons and eighteen battalions of mostly Austrian troops, and was entrusted with the role of containing the French left. Whilst Eugène was moving up, the enemy's guns opened fire, and to reduce casualties he ordered his men to lie down in their ranks and had Divine Service read at the head of each regiment. Meanwhile, the sappers toiled to build causeways over the Nebel and to repair a stone bridge demolished by retiring French outposts.

Infantry tactics at the beginning of the eighteenth century were broadly as follows. The foot of all armies fought in linear formations – in three ranks in the English army, in four or five in the French and many other Continental armies. The companies of a battalion extended away to left and right of their colonel (in the centre) in order of their commanders' seniority, whilst the grenadier company either took up the extreme right of the line or, as was the English practice, was divided into two platoons, one at each extremity.

The French fired their battalion volleys by single, double or treble ranks. Marlborough's infantry, by contrast, employed the platoon-firing system. By this, a battalion in line was sub-divided into eighteen equal platoons (heedless of companies), one-third of which fired together in a predetermined order, followed by the second and third 'firings', by which time the first group would have reloaded. Sometimes the fire of the front (kneeling) rank was reserved to form a fourth 'firing'. The advantages conferred by this system were continuous fire, better control – and therefore accuracy – and the fact that at any time one-third of the battalion would be loaded and prepared to deal with any sudden crisis. Marlborough did not introduce the system, which was probably of Swedish and Dutch origins, but he did insist on its adoption by all English battalions.

In his view of the cavalry's function in battle Marlborough was adamant that it should act as a shock-force, attacking directly with the sword. To enforce this view he would, as one observer wrote of him, 'allow the horse but three charges of powder and ball to each man for a campaign, and that only for guarding their horses when at grass, and not to be made use of in action'. Charges were delivered at a fast trot (*not* the gallop), usually by two squadrons, one forward, one in support 200 yards behind, each drawn up in line two ranks deep.

The French and several other nations tended to place greater stress on their cavalry as instruments of sophisticated fire-power; they would ride up, troop behind troop, to fire their pistols or carbines at the halt, before closing with swords for the mêlée. Marlborough's 'cold-steel' techniques proved far superior. In battle he invariably relied on cavalry reserves to deliver the *coup de grâce*, though he was at all times careful to provide his horsemen with infantry and artillery support in every sector of the field.

The Course of the Battle

At last, at 12.30 p.m., a messenger from the right reported that all was ready. Immediately Lord Cutts launched a series of fierce attacks in brigade strength against Blenheim. Two infantry onslaughts were repulsed (Brigadier Rowe being killed in the first), but attempts by part of the French cavalry to exploit the repulse were checked by well-directed platoon-fire. Indeed, so impressed was Clérambault by the fierceness of the onslaught that on his own authority he misguidedly summoned all of Tallard's infantry reserves until he had twenty-seven battalions and twelve dismounted squadrons packed into the village perimeter ('The men were so crowded in upon one another that they could never fire,' wrote one observer). This proved the fatal mistake of the day, but it went unnoticed by Tallard. He at the time was observing the Allied attack on Lutzingen and engaging in a furious altercation with his fellow-commanders.

From Oberglau, Tallard also watched the next incident near Blenheim, where at 1.45 p.m. five English squadrons led by Colonel Palmes were attacked by eight formations of the Gendarmerie, and proceeded to rout them. This comparatively minor event had a great effect on Tallard, and he later listed it as the first crucial failure leading to his eventual loss of the battle.

Marlborough, meanwhile, had ridden over in time to halt Cutts's impending third assault. Instead he gave Cutts new orders to contain the garrison of Blenheim, and thus secure the left flank of the Allied centre, which was about to cross the Nebel. By 2 p.m. Cutts had accomplished this: sixteen Allied battalions were now successfully bottling up twenty-seven French battalions. The Duke was then eager to achieve a similar success at Oberglau and so secure his right centre, and he rode over to hear how the Prince of Holstein-Beck was faring at the head of his seven-battalion attack on the village. To his consternation, Marlborough found that de Blainville had proved of tougher mettle than Clérambault, his colleague at Blenheim. At 2.15 p.m. he had sallied out with nine battalions (including the 'Wild Geese', units of Irish Catholic exiles in French pay) and badly drubbed his attackers. Two Allied battalions had been overwhelmed, and Holstein-Beck himself had been mortally wounded and taken prisoner.

This setback, together with a further development nearby, almost led to disaster. A few hundred yards to the south, the left flank of the Allied centre was just deploying from the Nebel marshes when it was violently attacked by part of Tallard's cavalry, which routed five squadrons of horse. The entire centre was now in jeopardy, but fortunately Marlborough – alerted by his aides – arrived in the nick of time to remedy both threatening situations. Three Hanoverian battalions were rushed forward to rally Holstein-Beck's dispirited troops, and Colonel Blood, commanding the artillery, by dint of Herculean efforts managed to move a battery across the Nebel. Its close-range fire checked de Blainville for the time being. In the meantime, the disciplined platoon-fire of the English battalions of the centre similarly repulsed Tallard's cavaliers, and soon Danish and Hanoverian squadrons under General Lumley arrived from the left, summoned by Marlborough, to complete the steadying of the line.

The crisis was not yet over, however, for now a new onslaught by part of Marsin's cavalry came sweeping up after passing in front of Oberglau, and crashed into the flank of the recovering Allied centre. Marlborough had no recourse but to send to Eugène for immediate aid.

far left Marshal Tallard, the Elector of Bavaria and their staffs survey the action in the centre. **left** Reeling from the heavy pressure of Marlborough's latest thrust, French cavalrymen in Tallard's centre break and flee for the Danube, where many drowned. **below** Tallard, taken prisoner by a Hessian regiment, is led up to Marlborough, who courteously places his coach at the French commander's disposal. Meanwhile a cow (left) is escorted to safer pastures.

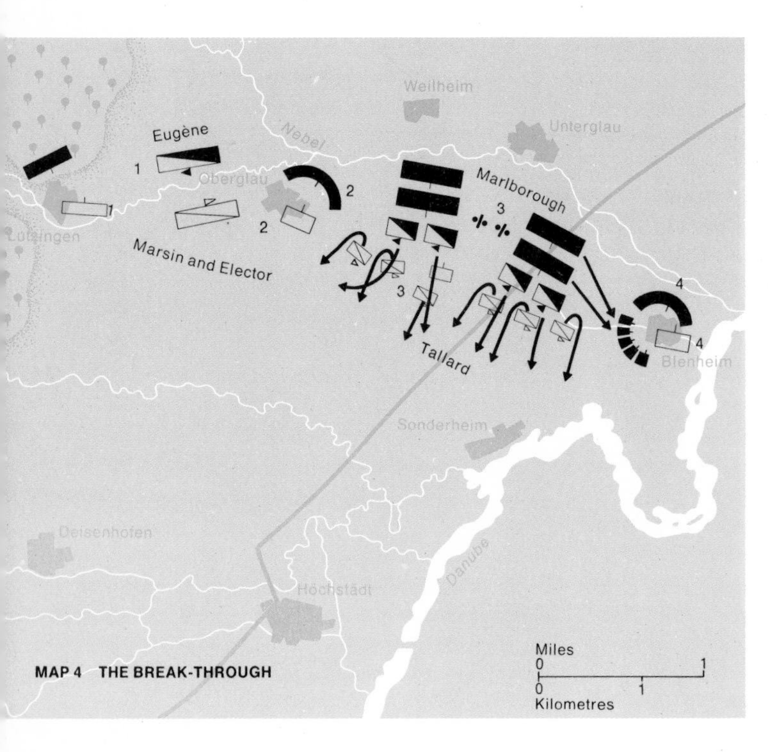

MAP 4 THE BREAK-THROUGH

By 5.30 p.m. all was ready for a final Allied attack in the centre. Marlborough's massed cavalry thundered up the slow incline at a fast trot and crashed into the blown and disordered remnants of Tallard's cavalry squadrons. In a moment they were broken and in full flight. The men of the nine infantry battalions fought bravely and died where they stood.

ALLIES

Infantry

Cavalry

Main batteries

1 Eugène
2 Holstein-Beck's infantry
3 Charles Churchill
4 Cutts

FRANCO-BAVARIANS

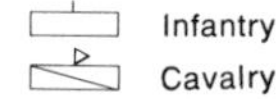
Infantry

Cavalry

1 Marsin and Elector
2 De Blainville
3 Zurlauben (9 bns + cavalry)
4 Clérambault

Prince Eugène, although hard-pressed himself on the whole of his sector, never hesitated. So great was his unselfishness and boundless trust in Marlborough's judgment, that he at once ordered General Fugger's brigade of Imperial cuirassiers – which constituted his sole reserve – to ride to the assistance of the centre. This force arrived at exactly the right moment to catch the near-victorious French, who were about to charge a second time. Piercing their flank, the Austrians proceeded to rout and decimate them. This event marked the real stabilization of the position in the centre, and by 3 p.m. Oberglau had been safely sealed off after de Blainville had been driven back into his defences. But it had been a close affair.

Although Eugène's wing had now fought itself to a near-standstill, in the centre the flanks were at last secured and the redeployment phase was almost completed. By 4 p.m. a total of eighty squadrons and twenty-two battalions was facing sixty and nine respectively of the French. The moment of decision had almost arrived. A lull settled over the whole battlefield, but by 4.30 p.m. the action had been renewed at all points. Away on the Allied right, the Danish battalions began to work round the Bavarian flank towards Lutzingen, whilst Eugène ordered Anhalt-Dessau to lead up the Prussian foot, and himself led the Austrian horse, in a renewed onslaught near Oberglau. In the centre, Marlborough gave the word for a general advance.

An apprehensive Tallard summoned the infantry of his right to support his cavalry, only to discover that none could respond to his call. In desperation he ordered his entire front line of cavalry to attack. The French caught the foremost Allied squadrons at a disadvantage, and flung them back in disarray. But help was at hand in the form of Major-General Lord Orkney at the head of his brigade. 'I marched with my battalions to sustain the horse, and found them repulsed, crying out for foot, being pressed by the Gendarmerie. I went to the head of several squadrons and got 'em to rally on my right and left, and brought up four pieces of cannon, and then charged.' The French horse drew off, discomfited, and the Allied centre fell back sixty paces to reform whilst Colonel Blood brought up a dozen guns which began to pour 'partridge-shot' (canister) into the nine French infantry battalions. By 5.30 p.m. all was again ready. Marlborough's massed cavalry thundered up the slow incline at a fast trot and crashed into the blown and disordered remnants of Tallard's cavalry squadrons. In a moment they were broken and in full flight. The nine infantry battalions fought bravely and died where they stood.

Tallard desperately tried to rally his horsemen in the camp area, but he was given no time to regroup. Part of the French cavalry fled towards Höchstädt; more headed for the Danube and its solitary pontoon bridge. Plunging down into the water-meadows near Sonderheim, they crowded towards the river bank, only to find the bridge broken under the press. Possibly 2,000 men drowned.

The French commander-in-chief rode for Blenheim village, but ran into a Hessian regiment which took him prisoner. He was led up to Marlborough, who courteously placed his coach at Tallard's disposal. Meanwhile, opposite Lutzingen, Eugène's third attack had failed to break the enemy line, although considerable ground had been gained. The Elector's and Marsin's thoughts at this juncture, however, were concentrated on one sole objective – escape. No attempt was made to re-establish any contact with the massed infantry within Blenheim; instead, the general of the left ordered an immediate retreat to Höchstädt, and by 7 p.m. they had successfully broken contact with the exhausted Imperialists. Blenheim was thus left to face its doom.

Shortly after that hour Charles Churchill had almost completed the full encirclement of Blenheim, covered by a new attack by Lord Cutts. The Allied army was understandably weary and in some disorder, so Marlborough decided not to interfere with the Elector's and Marsin's withdrawal, but to concentrate all his resources against Blenheim. Inside the village there were still 11,000 French infantry. It is true that they were leaderless – Clérambault having stolen away to find a watery grave in the Danube, whether by accident or design is not certain. But there was still a very real probability that the many fresh battalions might try to fight their way out through the weary encircling troops before the guns could be brought up in any number.

Orkney, however, was equal to the occasion. Noting the wind direction, at about 8 p.m. he set fire to some outlying ricks and buildings. 'The firing of the cottages, we could perceive, annoyed them very much, and seeing two brigades appear as if they intended to cut their way out through our troops, who were very fatigued, it came into my head to beat a parley, which they accepted of immediately, and their Brigadier de Nouville capitulated with me to be prisoners. . . .' Regiment after regiment followed suit, some burying their colours rather than see them taken. By nine o'clock the battle was over, and Marlborough had won 'the greatest and completest victory that has been gained these many ages'.

Soon Colonel Parke, Marlborough's aide-de-camp, was spurring for England, bearing the Duke's famous message to his wife scribbled on the back of an old tavern bill: 'I have no time to say more but to beg you will give my duty to the Queen and let her know that her army has had a glorious victory. Monsieur Tallard and two other generals are in my coach and I am following the rest. The bearer my Aide-de-Camp, Colonel Parke, will give her an account of what has passed. I shall do it in a day or two by another more at large.'

The Allied cavalry began to pursue the foe, whilst on the stricken field the grim business of the body-count began. It eventually transpired that Tallard had lost 20,000 killed and wounded besides a further 14,000 prisoners, including forty generals and 1,150 other officers. Sixty cannon, 300 colours and standards, and the entire camp fell into the victor's hands. The cost to the Allies was in the region of 12,000, including 654 officers and 8,029 in the Anglo-Dutch pay; of this number, almost a third had been killed.

Aftermath and Conclusion

By the Battle of Blenheim Vienna was saved, and thus also the Second Grand Alliance. At one stroke the Allies regained the strategic initiative and inflicted a severe blow to France's martial reputation, which would not be redeemed for at least another five years. The battle also demonstrated the superiority of English minor tactics, and of Marlborough's battlecraft. The immediate fruits of the victory were Ingolstadt (captured by a disgruntled Prince Lewis of Baden who resented his exclusion from the glorious day) and Ulm; before the campaign ended, moreover, an Allied army would capture Landau on the west bank of the Rhine. Although the war was still far from won, its structure had been completely altered, and England had emerged as a major military power for the first time since the Hundred Years' War some 250 years earlier.

As for Marlborough, Europe hailed a newly revealed and great commander. Honours were showered upon him. Queen Anne awarded him a pension of £5,000 a year and gave him Woodstock Manor, where Blenheim Palace was to be constructed at the nation's expense. The Emperor of Austria created him Prince of Mindelsheim – a small Bavarian enclave that had been recently acquired, together with the rest of Max Emmanuel's electoral possessions.

But Marlborough, who was for all his ambition an essentially modest and fair-minded man, was fully aware of the debt he owed to his comrade-in-arms, Prince Eugène, and above all to the officers and men of many nations who had served and fought so well under their orders. When, a few days after the battle, a woe-begone Tallard (whose son had been killed on 13 August) remarked that Marlborough had defeated 'the finest soldiers in the world', the Duke had the perfect reply ready. 'What then,' he asked, 'will the world say of the troops that beat them?'

Charles Grant at

LOBOSITZ

At Lobositz on the River Elbe the versatile bluecoats of Frederick the Great drove back an Austrian force commanded by Marshal von Browne. This was the first in a line of setbacks for the massed alliance of Austria, France and Russia that ranged itself against the Prussian King at the start of the Seven Years' War in Europe – an exhausting and bloody struggle in which Frederick's army proved itself to be the most formidable instrument of war yet seen on a European battlefield.

1756

Prussian infantry marching to take up battle positions near the Homolka.

The long series of wars which racked central Europe during the third quarter of the eighteenth century arose principally from the obsessive desire of Charles VI, the Holy Roman Emperor, to ensure the future integrity of his domains and to guarantee for them the succession he wished. His efforts resulted in the Pragmatic Sanction of 1713, whereby it was laid down that his daughter, Maria Theresa, should succeed him. But when he died in 1740, two of the most powerful of the rulers who had signified their assent to the Sanction, Charles Albert of Bavaria and Frederick of Prussia – who later came to be known as Frederick the Great – questioned the validity of the 1713 agreement.

Frederick promptly invaded the Austrian province of Silesia, so provoking the First Silesian War, or, as we know it better, the War of the Austrian Succession (1740–8). The Prussian king quickly fastened an iron grasp on this part of Austria, defeating the Austrians at the Battle of Mollwitz in 1741 – despite the fact that he himself left the field at the beginning of the action, leaving the battle to be won by his subordinate, Marshal Schwerin.

Of the several other states which came into the conflict, some did so officially and by formal declaration of war and some did so in a decidedly equivocal fashion, which, while it kept them ostensibly neutral, also permitted them to supply money and 'volunteers' to the combatant countries. France supported Bavaria; Great Britain, traditionally opposing France, favoured Maria Theresa – although, paradoxically enough, George II of England, in his capacity as Elector of Hanover, saw fit to declare his Electorate neutral.

In effect, Austria was isolated, and after a short time concluded an agreement with Frederick, by which he was left in possession of Silesia. Despite this agreement, however, Maria Theresa pursued her cause with great vigour and there was further fighting; Frederick defeated the Austrians at Chotusitz and Maria Theresa was obliged to cede Silesia for a second time. In the years that followed there were intermittent hostilities, and more than once Frederick came perilously close to annihilation. But his own determination and military skill saved him. Later, when he had restored his fortunes by several victories, he signed a treaty with Maria Theresa whereby she made over to him the greater part of Silesia. This treaty was signed at Dresden in 1745.

There followed a kind of stalemate during which the former combatants closely watched each other. Officially the War of the Austrian Succession ended in 1748 with the Peace of Aix-la-Chapelle. But the latter was an ambiguous agreement leaving much room for dissatisfaction; to the discerning eye there was little doubt that it was merely a question of time before there was a resumption of the fighting. It was left to Frederick, however, to light the fuse, albeit somewhat indirectly, when he signed the Convention of Westminster with Great Britain and Hanover in 1756.

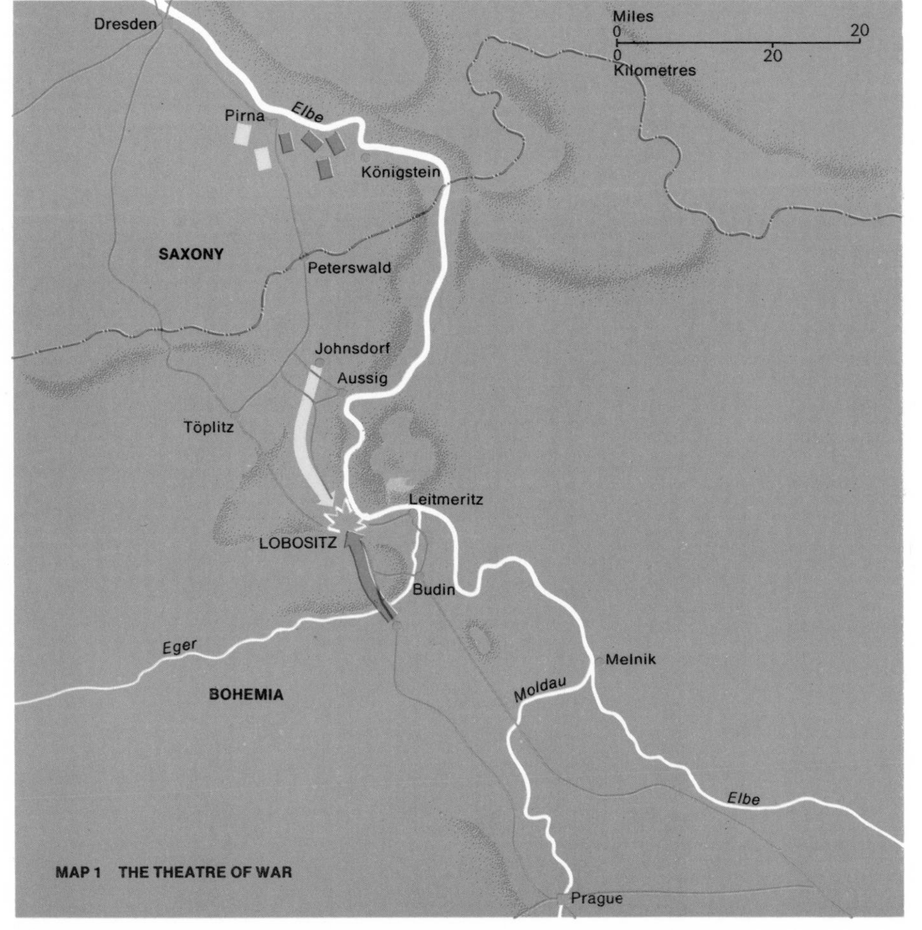

MAP 1 THE THEATRE OF WAR

On 29 August 1756 Frederick the Great of Prussia invaded Saxony from the north and took the capital, Dresden, in less than two weeks; the Saxon army withdrew to a heavily fortified region between Pirna and Königstein and Frederick crossed into northern Bohemia. At that point an Austrian army under Marshal von Browne moved up through Bohemia to Lobositz as part of a plan to link with the beleaguered Saxons. When he received news of this advance Frederick immediately led a Prussian force south from Johnsdorf to drive back the Austrians.

This effectively united France, Saxony and Austria against him, to which combination Russia added herself. Realizing possibly just how fraught his position was, and being informed that both Russian and Austrian troops were moving up to their frontiers with Prussia, Frederick took the bull by the horns and, on 29 August 1756, led an army across the border into Saxony. Less than two weeks later he occupied the Saxon capital, Dresden, and the small Saxon army of 20,000 men fell back to the south-east. News then reached Frederick that an Austrian army, 32,000 strong, under one of their most distinguished soldiers, Marshal von Browne, was advancing from Bohemia to effect a junction with the Saxons. To prevent this union, Frederick marched southwards with all speed and made contact with the Austrian army near the village of Lobositz (modern Lovosice), on the River Elbe.

The Rival Armies

Although records exist that tell us how the rival armies were organized, it is difficult to obtain a clear breakdown of numbers of infantry, cavalry, artillerymen, etc., at Lobositz. However, total strengths can be estimated with reasonable accuracy, and it is also known that the Prussians were superior in cavalry while the regular infantries of both sides were fairly evenly matched. Thus the comparative strengths of the rival armies may be summarized as follows:

Prussians

Infantry	22 battalions
Grenadiers	4½ battalions
Cuirassiers	41 squadrons
Dragoons	20 squadrons
Hussars	200 men

Total 30,000 men and 98 guns

Austrians

Infantry	30 battalions
Croat semi-regular infantry	4 battalions
Grenadiers	35 companies
Cuirassiers	48 companies
Carabiniers	8 companies
Dragoons	12 squadrons
Horse grenadiers	4 companies
Hussars	9 squadrons

Total 32,000 men and 94 guns

Marshal von Browne was already in Lobositz when his reconnaissance parties sent word that the Prussians were advancing on him from the north-west. At first there was some doubt as to whether the redoubtable Prussian king himself was with the approaching army; nevertheless the Austrian general saw that a major battle was in the offing.

Von Browne drew up his forces in an excellent defensive position. His right flank was firmly based on the River Elbe at Lobositz, and other troops were placed further

left and far left Austrian infantrymen from the period of the Seven Years' War, and an Austrian army mitre cap. **below left** An eighteenth-century Prussian bivouac, with its orderly lines of tents, men and wagons, guarded on all sides by outposts.

north along the river at Welhotta; the main part of the battle line stretched from Lobositz as far as the village of Tschischkowitz. Between there and Lobositz much of the Austrian army formed up to the rear of the Morellenbach, a network of marshes, ponds and small streams.

All this provided a first-rate natural defensive line through which even infantry, let alone cavalry, would have the utmost difficulty in passing. To strengthen his centre von Browne placed an advance guard, consisting of both horse and foot, in front of Lobositz, and a detachment of some 2,000 Croats – part of the semi-regular light infantry which abounded in the Austrian service – was posted well ahead on the Lobosch Hill. The latter was ideal ground for light-infantry action: the peak was covered with scrubby bushes and large boulders and was criss-crossed with ditches and stone walls. Across the valley from the Lobosch Hill – and also overlooking the road from Dresden along which the Prussians were coming – was another height, the Homolka. This was less considerable than the Lobosch, and consisted mainly of fairly gentle, grassy slopes. Between the main Austrian defensive line and the two hills lay a fairly level stretch of ground.

There had been some desultory firing between scouting parties during the night of 30 September, but this had amounted to little. However, by first light on 1 October the Prussians were very much in evidence. Long lines of dark-blue infantry were ranged across the valley between the Lobosch and the Homolka, and a great deal of artillery was in position all along the Prussian front. Behind the infantry could be observed dense masses of cavalry.

The troops that faced each other were basically similar: they were at the very zenith of an age of military professionalism. A century before, during the agonies and horrors of the Thirty Years' War, the civilian populace had suffered

Frederick II 'the Great', King of Prussia
Prussian Commander-in-chief
Frederick the Great (1712–86) was one of the dominant figures of the eighteenth century in Europe. During his early life – an unhappy time for him – military affairs were total anathema; he was far more interested in literature, and he cultivated and corresponded with some of the foremost writers of the time. He was forced into uniform by his father, Frederick William I, and court-martialled and disgraced when he attempted to escape. But when in 1740 he succeeded his father as King of Prussia, his entire character underwent a startling change. It seems that as soon as he had fully appreciated what a potent force his father had left him in the shape of the Prussian Army, he flung himself wholeheartedly into the military life, moulding an already well-trained body into the most formidable instrument of war yet seen on a European battlefield. After his accession Frederick lived, thought and dressed as a soldier. Sometimes his own previous successes in the field may have caused him to attempt the impossible – indeed most of his battles were fought against considerable odds – and he has often and with justification been accused of rashness. Nevertheless, with almost the whole of Europe ranged against him, his inflexible and unwavering determination never abandoned him and as a general he stands in the front rank.

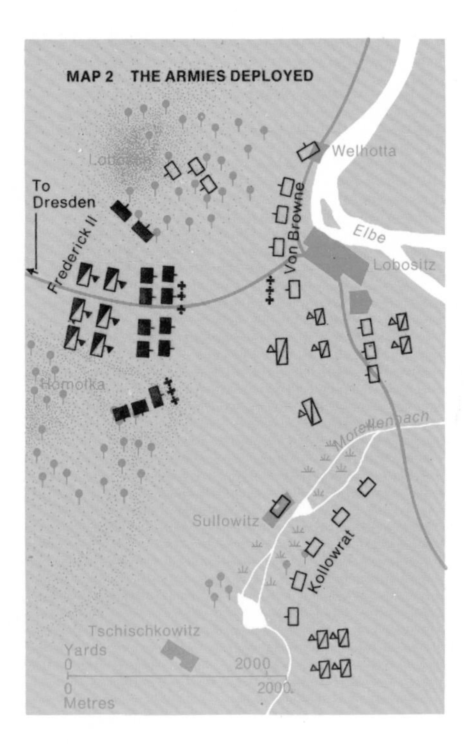

left Von Browne deployed his forces in a strong position, the right flank resting on the River Elbe, the centre and left extending between the villages of Lobositz and Tschischkowitz; in front of the main body of the army ran the Morellenbach, a watercourse that with the marshes surrounding it made an excellent natural defensive line. Other Austrian troops were placed on the Lobosch. Against the Austrians Frederick ranged his infantry and guns across the valley between the Lobosch and the Homolka; behind the infantry waited dense masses of cavalry.

AUSTRIANS
Infantry
Cavalry
Main batteries
PRUSSIANS
Infantry
Cavalry
Main batteries

right A view of the Austrian line moving up with artillery and cavalry support.

Marshal Maximilian von Browne
Austrian Commander-in-chief
Maximilian Ulysses von Browne (1705–57) was a comparatively unknown figure, although he had a substantial reputation in his adopted country of Austria. He was descended from the legendary Irish 'Wild Geese' – Catholic families that fled from Ireland at the end of the seventeenth century and whose members became mercenaries in almost every European country.

Von Browne rose to a high rank as a soldier in the service of Austria. He served with distinction against the Turks and in Italy, and was a subordinate commander against Frederick II of Prussia at the Battle of Mollwitz in 1741. For a great part of his service supreme command seems to have eluded von Browne and the present battle was one of the rare occasions when his authority was complete. It is possible that he was embittered by this, and that he felt that native Austrian commanders, especially those of princely stock, had been preferred to himself almost regardless of their qualities of leadership. He was nevertheless a most able and competent general officer, whose punctilious care for his men and insistence on being present where the fighting was thickest were both notable.

untold miseries of pillage, murder and rapine at the hands of the wide-ranging armies whose custom it was to live off the country. But in the third quarter of the eighteenth century the situation was vastly different. Armies were by then self-contained entities; they dragged enormous baggage trains wherever they went, and consequently their disruptive effect on local populations was to some extent decreased.

In basic military tactics, too, there had been important changes. Generally speaking, the highest degree of training was to be found in the infantry, and in this field the Prussian soldier was a model for all others. Drilled to the last degree, he could load and fire his musket at the then rapid rate of four times a minute (in this he was helped by the recent adoption of the metal ramrod instead of the old wooden one, which had readily snapped under the stresses occasioned by rapid fire). The volleys of the Prussian infantrymen, although ineffective at long range, were murderous when the closely ranked enemy came within forty or fifty yards. At that time hand-to-hand combat was rare among the infantry: long before the two sides could reach close quarters one almost invariably broke before the sustained fire of the other.

As to the other branches of the army, Frederick himself was the first to operate a rudimentary species of horse artillery, in which all personnel were mounted, thus giving the guns far greater mobility. At the same time, too, light

troops, both horse and foot, were finding a permanent place in almost every army. At the beginning of the 1740s the Austrian cavalry had tended to dominate the battlefields on which they fought. But, thanks to Frederick's intensive programme of training and the inspiration of great cavalry leaders such as von Seydlitz, the Prussian squadrons were rapidly improving in manoeuvrability and striking power. Even so, the musket was the supreme weapon and it was becoming increasingly difficult for even the heaviest cavalry to penetrate the ranks of infantry which was properly formed and prepared. The volume of fire and the lines of gleaming bayonets were usually sufficient to force the horsemen to veer away. Only when the cavalry was able to deliver a surprise attack on the flanks of unprepared foot – as von Seydlitz did at Rossbach against the French – was a cavalry charge really effective.

Those, broadly, were the chief elements of warfare in the mid-eighteenth century. Battles were fought by solid blocks of infantry, flanked by squadrons of cavalry, with small groups of light infantry skirmishing ahead and falling back as the main bodies closed with each other, while batteries of artillery, posted on the flanks to give enfilading fire or on adjacent heights to enable them to fire over their own troops, blazed away at the enemy with roundshot or shell.

The Course of the Battle

The first action took place on the Lobosch, when strong bodies of Prussian infantry pressed up the slopes, forcing their way past the walls and other obstacles and driving back the Austrian skirmishers. At the same time a powerful battery of Prussian cannon on the Homolka came into action, delivering concentrated and effective fire. The Austrian artillery, however, was not slow to reply and its guns took considerable toll of the Prussians, including two of Frederick's generals who were mortally wounded during the early discharges. Several Austrian infantry units were brought forward to reinforce the arc of troops in front of Lobositz, where Marshal von Browne took personal command, leaving General Kollowrat to assume responsi-

Prussian cavalry – part of a strong force totalling sixteen squadrons – riding past its own infantry to muster south of the Homolka before charging the Austrian lines.

bility for the bulk of the infantry force posted behind the Morellenbach.

Shortly after 8 a.m. the sun was shining but there still remained some early-morning mist in the lower-lying parts of the battlefield; this was thickest, naturally enough, along the course of the Morellenbach. From his position on the Homolka Frederick could see a considerable number of Austrian troops beyond the stream but, with the persistent mist, it was far from easy to determine either their composition or their strength. However, despite the obvious possibility of flanking fire from the Austrians lining the stream, Frederick brought up a strong force of cavalry – sixteen squadrons – alongside the right wing of his infantry and mustered it just south of the Homolka. His plan, it seems, was for the cavalry to advance straight ahead and drive off the Austrian horse who were to be seen protecting the left of Lobositz.

The Prussians went forward in an impressive mass, their left flank towards Sullowitz, a village lying some little way in front of the Morellenbach. Austrian infantry were ensconced there and their sudden fire surprised and disconcerted the advancing Prussian cavalrymen, who suffered considerable losses. Smitten by this fire, they began involuntarily to swerve away to their left to avoid the musketry. This move instituted a swirling cavalry engagement, for the Prussians, as they changed direction, were charged on their right flank by several squadrons of Austrian dragoons (these being under the command of one O'Donnell, another representative of the 'Wild Geese').

However, the latter were themselves flung brusquely back by a second wave of Prussian cavalry, and the whole seething and heaving welter of horsemen surged towards the Austrian lines, hewing and hacking at each other in great confusion. Light-infantrymen in their path scattered to avoid being crushed by the horses' hooves, and two fresh Austrian heavy-cavalry regiments came up with a rush to assist their comrades. Their headlong charge was sufficient to sway the fight in favour of the Austrians, and Frederick's cavalry went thundering back towards its own lines.

Meanwhile a vigorous infantry combat was raging over the Lobosch. Fresh Prussian infantry – altogether eleven battalions, possibly as many as 5,000 men – had joined the original attackers and the most murderous sort of close-range fighting was going on, the infantrymen using the cover of walls and hedges to advance in short rushes. As they did so the heavy musket balls from the Austrian side sought them out and left the slopes littered with blue-clad bodies. Strangely enough, despite their classic discipline and their normal adherence to the tenets of the close-order advance, the Prussian foot adapted well to this new kind of light-infantry work and were soon proving themselves more than a match for the Austrian Croats; the latter were eventually so hard pressed that von Browne sent forward reinforcements, including a number of his élite companies.

The initial repulse suffered by Prussian cavalrymen had failed to daunt either their spirits or those of their comrades who had yet to fight. Frederick, it seems, had not issued orders for a counter-attack when the Prussian cavalrymen who had been held back behind the infantry during the first charge – forty-three squadrons in all – suddenly poured forward through the gaps in the infantry line. Having executed this most unexpected manoeuvre, they formed up and went galloping forwards in a most tremendous charge against the Austrians. Not only the follow-up units but those which had already been in the fighting joined this wild surge of upwards of 10,000 horsemen across the flat ground in front of the Morellenbach.

There can have been few occasions in the numerous wars of the eighteenth century when greater numbers of cavalry were engaged in such a mêlée, much of which, to add to the confusion, was obscured by clouds of smoke from the incessant discharge of cannon and musket. The flanking fire from Sullowitz was this time ineffectual and the whole mass crashed at the gallop into the Austrian cavalry. It is

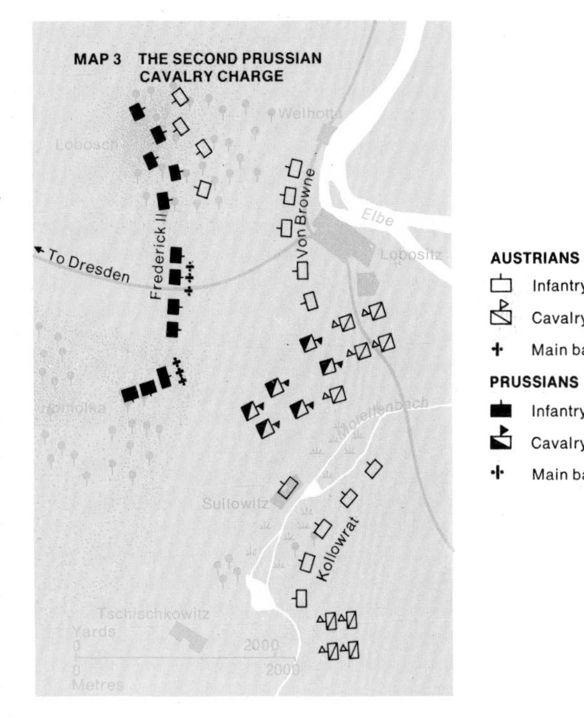

This spectacular if ill-disciplined manoeuvre had some 10,000 Prussian cavalry surging wildly against the Austrian lines; such was the scale and impetus of the charge that the mêlée with von Browne's cavalry carried deep into the Austrian defences. But once the Prussian hordes had been stopped, von Browne attacked their flank with two regiments of cuirassiers and successfully put them to flight.

not clear whether the Austrians stood fast to receive the onslaught – it would probably have been unwise to do so – or whether they attempted to counter-charge, but in any event they were flung back by the sheer weight and impetus of the Prussians. Nevertheless the Austrians fought with the greatest determination and managed to bring their assailants almost to a halt some way to the south-west of Lobositz. As we have already noted, Marshal von Browne was in personal command of this section of the line, and he seems to have been keenly aware of the critical nature of the situation. He reacted promptly, bringing up two regiments of cuirassiers (armoured cavalrymen) from his reserve and dashing them against the flank of the Prussian horsemen, who by this time were milling about in some disorder, impeded by their large numbers from manoeuvring in effective concert.

Frederick's cavalrymen reeled under this new charge but replied lustily with their long sabres. However, the heavy Austrian troops proved difficult to cope with, and the Prussians, feeling also the effects of renewed pressure from the reforming Austrian regiments to their front, began to fall back. After some ten to fifteen minutes the withdrawal became a rout and the Prussian horsemen dashed panic-stricken to their rear. Many were made prisoner by their jubilant pursuers, and General von Seydlitz himself had to be rescued from possible capture by a group of hussars. The Prussian cavalrymen who were able to make good their

above and left The beginning of the first cavalry mêlée as seen from behind the Prussian position. **below** General von Seydlitz is shown a group of captured Austrian flags following the first charge.

escape retired behind the steady line of their infantry and remained there, unable to take further part in the fighting.

All this time heavy Prussian guns had been hurling roundshot against the Austrian lines – and Marshal von Browne had more than one horse shot from under him. But his men held their position. Even on the bloodstained slopes of the Lobosch the Prussians were making only slow headway and shortly after noon it seemed that the Austrians held a slight advantage, little or no impression having been made by the enemy on their defensive position. However, the powerful reinforcements that Frederick had pushed up to assist the troops attacking the Lobosch were beginning to make their presence felt, and at a critical moment the officer commanding the Austrians on the hill, General Lacy, fell wounded. As is not unusual in such circumstances there was some little delay before the chain of command could be re-established, and in that period of indecision the Prussian infantry came pouring in great numbers down the eastern side of the hill, driving before them the remnants of the Croat battalions.

This was a time of considerable danger for the entire Austrian right wing, with the Prussians pushing boldly forward towards both Welhotta and Lobositz. Troops in the former village were hastily withdrawn, heavy guns in danger of being captured were moved to safety, and fresh Austrian infantrymen came up at the double to steady the crumbling position. Their arrival was timely, and the situation was to some extent restored, although the loss of the Losbosch was evidently a serious blow to von Browne's defensive plan.

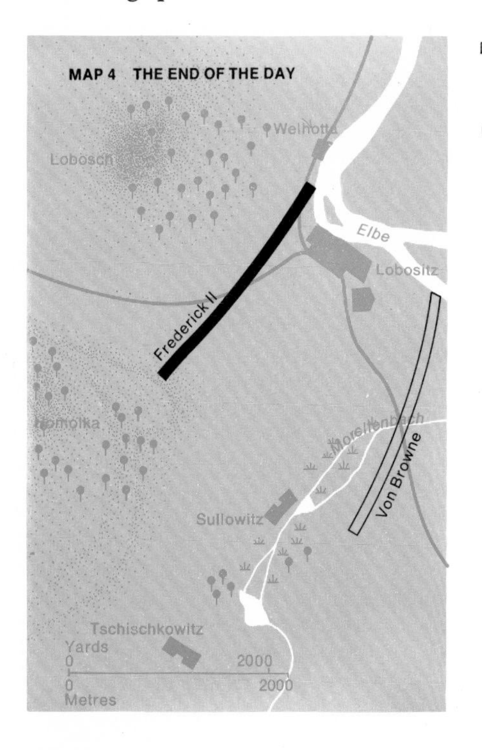

Final Austrian position

Final Prussian position

The battle ended shortly after the Austrians had completed a masterly fighting withdrawal from the burning village of Lobositz. The Prussians were then barred from further pursuit by the main part of the Austrian army, which moved across and took up a secure position to the rear of the village, with its right flank resting squarely on the River Elbe.

Lobositz itself was now under heavy attack from the Prussian infantry, which pressed with great boldness as far as the outskirts of the village, much of which was now in flames after being shelled by Prussian howitzers; hand-to-hand fighting then broke out around the outer fringe of houses. Although they defended the outskirts of Lobositz with great vigour, the Austrians were acutely aware that the conflagration in the main part of the village was rapidly increasing and would soon cut off their retreat. Under von Browne's supervision, a fighting withdrawal was ordered. This was carried out impeccably, numbers of cavalry being brought forward to cover the operation. To conform with this manoeuvre a backward wheeling movement was ordered along the entire line and the whole Austrian army swung back, closing up tightly to the River Elbe so that its right flank rested squarely on the river.

Since the great cavalry fight earlier in the day the western section of the battlefield had been practically without incident save for continuous artillery fire from both sides. Now, following this latest withdrawal, the fighting almost completely died away and by the middle of the afternoon all was quiet, except for the groans of the wounded. The battle had been a hard-fought one, and casualties amounted to around 3,000 on each side. Frederick could consider himself more than a little fortunate: he, after all, had attacked an almost equal force in a strong position – circumstances in which the attacker would normally expect to sustain heavier losses.

Soon, torrential rain began to turn the battlefield into a near-swamp. Neither of the two commanders was clear what the other might do next, but von Browne, whose supply problem had become acute owing to the disappearance during the fighting of much of his baggage train, decided that there was little point in maintaining his presence in the Lobositz position. Consequently he ordered a withdrawal, ostensibly to put his army into a better position in which to help the Saxon forces. On the Prussian side Frederick was unsure of the best course to take, and even contemplated withdrawal himself, but promptly decided against this measure when he was informed of the Austrian retreat. Instead he took the opportunity to claim a resounding victory, in which he had destroyed many thousands of Austrians.

Aftermath and Conclusion

After Lobositz Frederick pressed northward and received the surrender of the Saxon army at Pirna on 16 October. But the political alliances formed against him were wide-ranging and powerful, and Frederick's Prussians could ill afford to lay down their arms. Between 1757–62 many great armies marched, countermarched and fought many a bloody battle across central Europe until sheer exhaustion dictated that there should be peace. Armistices were eventually signed by all the great powers, and Austria, Russia and Prussia ultimately came to terms with each other. Frederick, happily for him, remained in possession of Silesia, the rich morsel he had coveted for so many years and which, to a large extent, had been the direct cause of the war.

As for Frederick's opponent at Lobositz, Marshal von Browne continued to serve his Empress well – but for only a few months longer. Despite the poor state of his health at the time, he commanded part of the Austrian army under Prince Charles at Prague in the summer of 1757 when it was attacked by large Prussian forces. Fighting raged for several days, and it was while he was engaged in spirited and heroic efforts to direct his men that von Browne was shot and mortally wounded. He lingered on in great pain for weeks, dying on 16 June 1757.

Compared with some of the later battles of the Seven Years' War, Lobositz was an indecisive encounter. However, for Frederick it paved the way for the surrender of the Saxons, and for both sides it provided valuable lessons in the war that followed. Lobositz also gave a lively indication of the quality of the Prussian infantry and of its ability to fight not only in the conventional close order but – as took place on the Lobosch – in extended order and in the role traditionally associated with light infantry. Little wonder is it that Frederick's bluecoats dominated the battlefields of Europe for as long as they did.

Aram Bakshian Jr at

SARATOGA

At a critical moment in the American Revolution General Burgoyne's British army, arrayed as though to fight a stately set-piece in the European manner, was trapped in the New England forest by American rebel forces. Near Saratoga, after two ferocious battles with General Gates's American army – at Freeman's Farm and Bemis Heights – Burgoyne was forced to surrender. This act decisively changed the course of the Revolution.

1777

Action in the clearing at Freeman's Farm during the First Battle of Saratoga. This was the scene of a bloody struggle that lasted for three hours, in which Burgoyne's orthodox formations narrowly held their own against the American sharpshooters and waves of reserves brought forward by the tigerish Benedict Arnold.

The Background to the Campaign

From the beginning of the American Revolution in 1775, British and Loyalist leaders harboured the hope that if New England, which they considered the nerve-centre of the rebellion, could be isolated and subdued, order could quickly be restored in the central and southern colonies. When Major-General Sir Guy Carleton, the Governor of Quebec, successfully repulsed a makeshift American invasion in 1776, the prospect of a British drive from Canada, severing New England's communications with the rest of the continent, became a real one. Only a hard-fought naval action on Lake Champlain, in which a hastily improvised rebel fleet under General Benedict Arnold (temporarily made naval commander) inflicted heavy losses on Carleton before being destroyed, postponed the British invasion until 1777.

The man who would lead the British invasion was not, however, Sir Guy Carleton. Though one of the ablest officers to serve the Crown during the Revolution, Sir Guy had at least one rival with superior influence in London. That man was Lieutenant-General John Burgoyne, who, after observing the war as an inactive subordinate in 1775–6, returned to London to submit a document entitled 'Thoughts for Conducting the War from the Side of Canada' to Lord George Germain, George III's incompetent Secretary for America. In the plan Burgoyne, eager to secure his own independent command, proposed a grand northern strategy; and he used his considerable literary flair to make it seem almost foolproof. He called for a three-pronged attack which would, he claimed, end the costly American war within a year. It was implemented by the Ministry in the following form:

1. A major northern army, commanded by Burgoyne, was ordered to invade the area that is now New York State from Canada, proceeding southward along Lake Champlain and the Hudson River.
2. A smaller force, commanded by Colonel St Leger, would simultaneously invade from Oswego by way of the Mohawk Valley, supposedly an area of concentrated Loyalist support.
3. The third prong of the attack would be the main force, consisting of a strong detachment of the central British army in the colonies, commanded by Lieutenant-General Sir William Howe and based in New York City. It would be dispatched north up the Hudson to join the other two columns in Albany, thereby slicing through the spinal cord of the rebellious colonies.

It was an imposing plan – on paper. Yet it was based on a number of questionable assumptions. Thanks to over-optimistic advice from Loyalist emigrés in London and Canada, the British were led to believe that thousands of well-disposed New Yorkers would rally to the invading forces as they advanced on Albany. This never occurred. In fact, after some ugly incidents involving Burgoyne's Indian auxiliaries and civilian victims, thousands of previously neutral or apathetic colonists took up arms with the rebel cause.

Problems of logistics and geography were not treated realistically. The Burgoyne plan sadly underestimated the difficulties that even a small band of ill-treated enemies could cause a conventional army, overburdened with heavy infantry and artillery – not to speak of officers' baggage – as it marched along a forest route.

Yet the plan did have tremendous potential for crushing rebel morale if it could be implemented quickly, while the enemy still reeled under the impact of past defeats, and no adequate American army blocked the way between Canada and Albany. The success of Burgoyne's thrust relied

In his plan for subduing the rebellious Americans, Burgoyne proposed a three-pronged attack. He himself would invade from Canada, penetrating as far as Albany where he would unite with a smaller force, led by Colonel St Leger, moving in from Oswego, and with a third, more powerful body from New York, commanded by Lieutenant-General Sir William Howe.

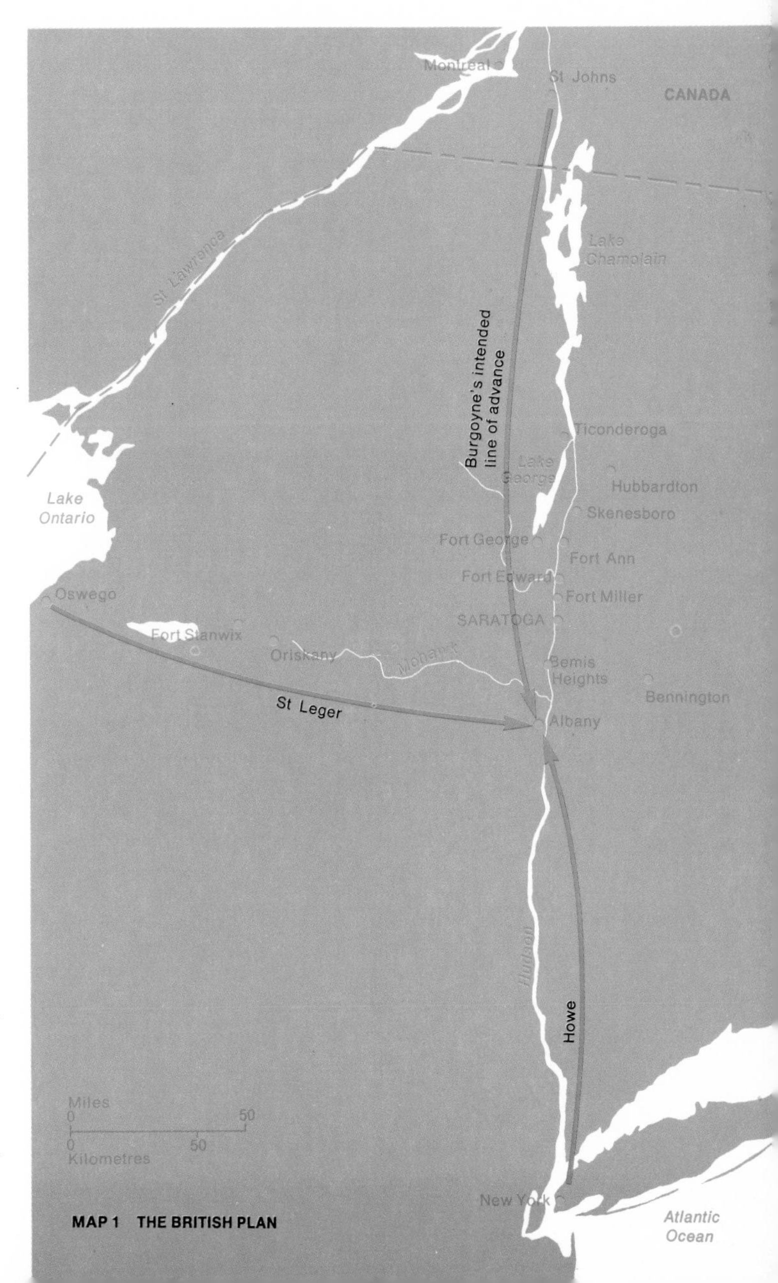

MAP 1 THE BRITISH PLAN

mainly on speed and shock – on striking home before a stunned enemy could react decisively.

Finally, and most importantly, it could only succeed if Burgoyne, St Leger and Howe operated in the closest concert. This, in turn, required firm and consistent orders from London and open, prompt communications among the three commanders. Instead, Howe was later authorized by Germain to follow his own, conflicting plan of campaign, driving south into Pennsylvania with his main army and leaving his subordinate, Major-General Sir Henry Clinton, to make shift in New York City with a drastically reduced force. As for communications, they were irregular and subject to delay and interception almost from the outset, and grew more unreliable as Burgoyne and St Leger penetrated deeper into enemy territory.

The Campaign

None of these potential difficulties weighed heavily on General Burgoyne on 17 June 1777 when he embarked from St Johns, Quebec, with a heterogeneous army of some 9,500 British regulars, German mercenaries from Brunswick and Hesse-Hanau, Canadian irregulars and Indian auxiliaries. His first target was Fort Ticonderoga, a ramshackle pile sometimes grandiosely referred to as the 'Gibraltar' of the North American continent.

Awaiting him at Fort Ticonderoga was a demoralized, ill-equipped force of 3,000 rebels under General Arthur St Clair, a Scot who had once held a commission in the British Army. St Clair's lines had been designed for a defending force of 10,000, and his alarm may be imagined when Burgoyne's splendid flotilla disgorged its invading masses. The British gunners, goaded by Major-General William Phillips, Burgoyne's second-in-command and a former artilleryman, dragged cannon up the steep slope of Mount Defiance, a craggy hill overlooking the fort, which the Americans had neglected to fortify. By 5 July the batteries were ready and the situation for St Clair's garrison was extremely dangerous. To his credit, St Clair decided to abandon the post. Although he must have known that he would bring down a storm of censure from extreme elements in the Continental Congress, and perhaps personal disgrace, he also knew that a quick retreat was the only way of saving what was then the main regular American army in the Northern Department. Under cover of night he retreated to the eastern shore of Lake Champlain over a pontoon bridge.

Ticonderoga had fallen virtually without resistance, but the American garrison had saved itself to fight another day. Burgoyne, who pursued that part of the American force which had fled by water – mainly the wounded, support troops and transport – met with little resistance and was able to capture important supplies and destroy what remained of the American lake fleet. The main American column, retreating overland, was hotly pursued by

American soldiers from the Revolutionary War, by H. A. Ogden. In the foreground is one of Morgan's Riflemen, a member of an élite unit of marksmen. Unlike the bulk of American fighting men in the war, who used the same kind of smooth-bore muskets as their British and German adversaries, Morgan's and certain other specialist units used the American long rifle, which had a considerably greater range. Whereas it was almost impossible to hit an individual target beyond 150 yards with a standard musket (though unaimed, compact volleys could still take a toll on mass targets), a skilled rifleman could with ease hit a seven-inch target at 250 yards, and ranges of 3–400 yards were not unknown.

Brigadier-General Simon Fraser, Burgoyne's tireless and intelligent commander of light troops. He overtook the American rearguard on the morning of 7 July near Hubbardton, Vermont. After heavy fighting, the American force was scattered. The next day, near Fort Ann, the British routed another body of Americans.

At a cost of nearly two hundred casualties, John Burgoyne had successfully achieved what he originally expected would be the most difficult phase of the invasion. Unfortunately for him, this unlooked-for success prompted him to make a critical error. Rather than return to Ticonderoga (he was reluctant to make a retrogressive move on the crest of victory) and then effortlessly advance down Lake George by water, Burgoyne determined to proceed overland. Only sixty miles lay between him and his goal of Albany – but each would be a nail in the coffin of his plan for a speedy conquest of the rebellious colonies.

Contemplating Burgoyne's army as it inched painfully along the forest route to Albany, General Nathaniel Greene, one of George Washington's ablest lieutenants, predicted that 'General Burgoyne's triumphs and little advantages may serve to bait his vanity and lead him on to his final overthrow'.

Twenty-three of the sixty miles lay between Skenesborough, where the march began, and Fort Edward. It was a rough path through dense forests, marshes and small rivers, and the natural obstacles were reinforced by a corps

left A British grenadier, *c.* 1780. British infantry battalions were at the time divided into ten companies. The two flank companies, grenadiers and light infantry, were often detached to form composite specialist battalions within brigades or armies. **below** Burgoyne addressing a group of Indians; at Saratoga about 100 Indian auxiliaries served on the British side.

of American woodsmen under orders to fell trees all along the British line of march. These orders, issued by General Philip Schuyler, then commander of the Northern Department, played as important a part in sealing Burgoyne's fate as any decision made on the field of battle. Thanks to them, it took the British army twenty-four days to travel the twenty-three miles to Fort Edward.

As the surrounding countryside was swept clean of provisions and crops by the retreating rebels, supplies became increasingly scarce, and, with each painstaking mile, Burgoyne's line of communications became more tenuous and costly; to maintain them, valuable troops were continually siphoned off from the main force. By 16 September his army could boast of no more than seven additional miles' progress down the shore of the Hudson to an abandoned trading post known as Fort Miller. The shortage of horses and provisions (an entire regiment of Brunswick Dragoons had had to stumble through the campaign on foot, the laughingstock of the army) added to the difficulties of the British commander.

Meanwhile, the second and third prongs of the Burgoyne plan were being blunted or destroyed. On 6 August St Leger's force, a motley assemblage of just over 1,500 British, Tories (Loyalists) and Indians, had engaged in a small but fierce encounter with American militia at Oriskany – one of the bloodiest engagements of the war. Though victorious, St Leger's little army never completely recovered and, little more than two weeks later, while besieging Fort Stanwix, the whole body retreated in considerable disorder at the approach of a large relief column under the fiery Benedict Arnold. From that moment the second prong ceased to exist.

As for the third force, this had been thoroughly blunted when Sir William Howe, with the strange acquiescence of Germain in London, had transported most of his army south to capture the rebel capital of Philadelphia, a prestigious feat but of little real value to the Royal cause.

Strategically, the die had been cast. Tactically, the situation was still fluid, but time was now on the Americans' side. Even so, it was possible for Burgoyne to retrieve the

situation if he could replenish his supplies, secure adequate horses, and win a quick, crushing tactical victory over the American Northern Army. Alternatively, he still had the option of retreating with the bulk of his force intact – abandoning his grand design and risking his future career, but saving his army.

Burgoyne opted for the first alternative, but carried out his plan in the worst possible way. He sent out a foraging expedition of 800 men, including the hapless, horseless Brunswick Dragoons, under the command of their colonel, Friedrich Baum – an officer who, whatever his other virtues, had little knowledge of irregular warfare and even less of the English language. Although the Indians accompanying Baum confined their hostilities to slaughtering the livestock captured along the way, their mere presence served to inflame the local farmers and, on 16 August at Bennington, Vermont, the expedition was surrounded and destroyed by a superior force of American militia. A relief column belatedly sent out by Burgoyne and consisting mainly of slow-moving Brunswick grenadiers, bogged down with superfluous gear and looking strangely out of place in the middle of the American wilderness with their jutting moustaches and tall brass mitre caps, was sharply engaged and driven back with serious casualties.

Instead of providing fresh supplies, morale and momentum, the raid into Vermont had cost Burgoyne 800 men, four light field pieces, and, perhaps the greatest loss, his army's faith in its own invincibility. Writing on 22 August George Washington, Commander-in-chief of the Continental army, urged his fellow colonists on to the kill. 'Now,' he wrote, 'let all New England turn out and crush Burgoyne.'

On the other hand, the rebel cause in the north had received the first important boost to its morale in the campaign. This, together with stories of Indian atrocities (both real and imagined) which were spread by American propagandists, brought in wave after wave of fresh recruits to block Burgoyne's advance to Albany.

Too late Burgoyne began to grasp the seriousness of his plight. 'Wherever the King's forces point,' he complained, 'militia to the amount of three or four thousand assemble in twenty-four hours.' Writing to Germain in London, he stated that it might be best for his army to stay put, or actually withdraw 'had I latitude in my orders'. It was a curious remark, with the faint air of excuse about it, since John Burgoyne's thinking had had no small influence on the drafting of the orders in the first place. However, he continued his advance, which now led to the two battles of Saratoga. His move was the last fling of an inveterate gambler with a blind faith that, in the end, his difficulties would simply evaporate. Gentleman Johnny, as vain as he was charming, preferred precarious muddling through to the personal humiliation of an early retreat. On 13 September his army crossed the Hudson River at Saratoga and severed its lifeline to Canada.

The First Battle of Saratoga (Freeman's Farm), 19 September 1777

The American army that awaited Burgoyne at Bemis Heights had grown from 3,000 to 7,000 men. Through political intrigue, General Schuyler had been replaced as

left Burgoyne's troops edge their way along the forest route to Albany. After the taking of Ticonderoga, Burgoyne made his first major error in choosing a cross-country route for his ponderous army; ill-equipped for the rough terrain, many of his men fell to rebel snipers, and, at Bennington, an entire foraging force of 800 men was surrounded and destroyed.

In reality Burgoyne's imposing invasion plans failed in every major respect. His own army, the northern force, received several rude shocks before its eventual isolation and defeat at Saratoga; meanwhile St Leger's force was halted at Fort Stanwix, and Sir William Howe transported most of his army south from New York to attack Philadelphia.

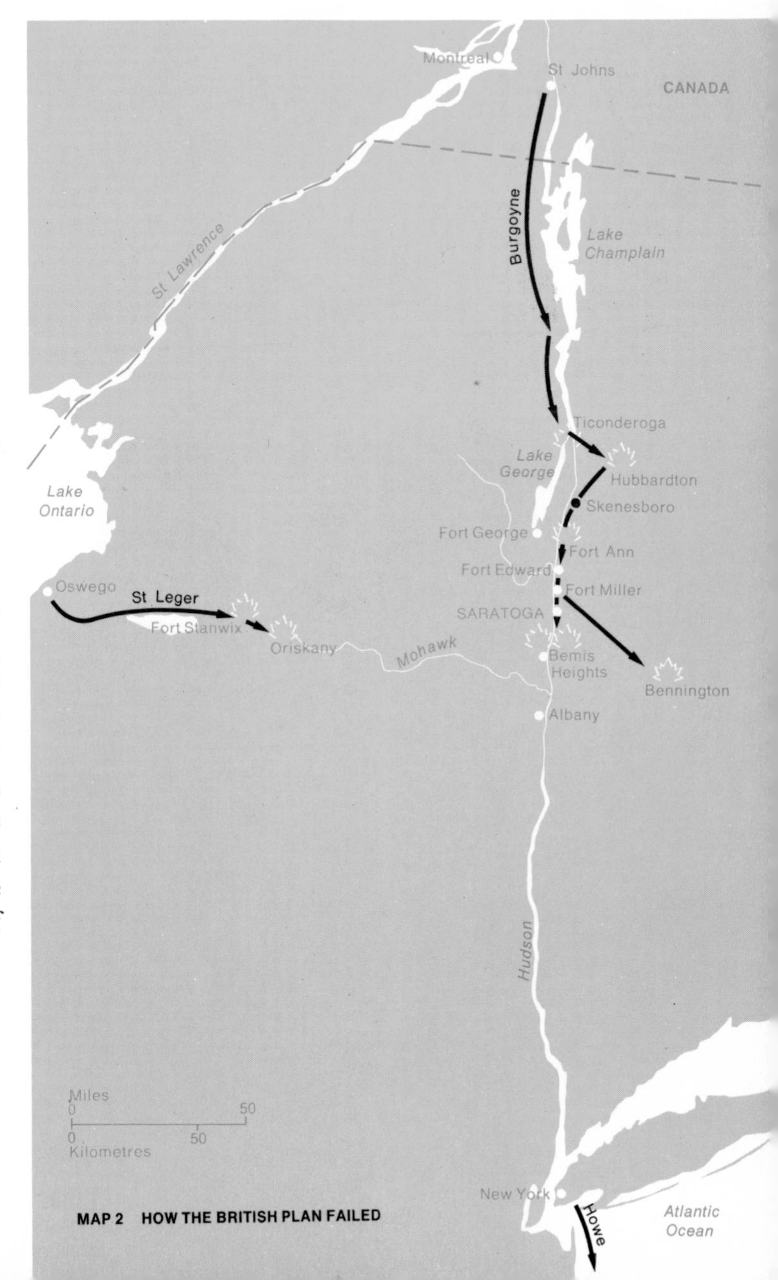

MAP 2 HOW THE BRITISH PLAN FAILED

commander by Major-General Horatio Gates on 19 August.

Doubting the ability of his men to face up to British regulars in pitched battle, Gates had concentrated on fortifying his position – and it would have been difficult to find a better one. Colonel Thaddeus Kosciuszko, a Polish volunteer and military engineer (he later led his own country's unsuccessful struggle for independence in Europe) had constructed a strong defensive line studded with batteries along the clear bluffs overlooking the forest to the west and the narrow plain along the bank of the Hudson to the east. Most of the line was reinforced by breastworks of fallen trees that were restrengthened in the period between the First and Second Battles of Saratoga.

The comparative strengths of the rival armies at the First Battle of Saratoga were approximately as follows:

Americans	**British**	
7,000 men and 22 guns	Right-hand column (Fraser)	2,500
	Centre (Hamilton)	1,400
	Left (von Riedesel)	1,800
	Total 5,700 men and 35 guns	

Gates posted three brigades of tested Continental regulars (those of Nixon, Patterson and Glover) under his personal command on the right wing of the works. In the centre and left he posted two more Continental brigades

Lieutenant-General John Burgoyne
British Commander-in-chief
'Gentleman Johnny', as General Burgoyne (1722–92) was affectionately known to his men, was fifty-five years old when he faced the rebel army at Saratoga, and an intelligent, experienced soldier with a reputation for gallantry that extended into his private life. He was a humane commander, idolized by all who served under him and sincerely interested in the welfare of his troops. A pioneer of light cavalry in the British Army, he performed some dashing raids on the Iberian Peninsula during the last phase of the Seven Years' War.

Possessed of considerable talents, Burgoyne was inclined to spread them a bit thin. He dabbled in politics (as a Whig MP) and literature (primarily comedy-writing for the London stage), and was something of a society figure, having many powerful connections at Court and at Westminster.

Major-General Horatio Gates
American Commander-in-chief
Born in Maldon, England, Horatio Gates (1727–1806) was the son of a housekeeper in the service of the Duke of Leeds. His father, a minor government official, managed to purchase young Horatio a commission and, at the age of twenty-two, Gates left England to serve in a variety of places, including Nova Scotia, the American wilderness (with the ill-starred General Braddock) and the West Indies. In 1762, at the age of thirty-four, he was commissioned a major, but the end of the Seven Years' War found him back in England, an idle, embittered half-pay officer. Never fully accepted by his social superiors, Gates became something of a political radical, and in 1772 he emigrated to America, settling on a plantation that he had purchased in Virginia. With the outbreak of the Revolution, he emerged from obscurity to become the first adjutant-general of the newly formed Continental Army, forsaking all loyalty to the Crown. His energetic, efficient staff work soon earned him a following among influential members of the Continental Congress. Gates's talents were those of an administrator; his great failings were his personal pettiness and a natural timidity aggravated by his lack of combat-command experience.

(Poor's and Learned's) and a crack 330-man corps of Dan Morgan's Riflemen – weather-beaten backwoodsmen in hunting shirts, and by far the keenest marksmen in America. Together with 200 light infantrymen under Major Henry Dearborn, they formed an élite body of sharpshooters unmatched by any that Burgoyne commanded. So highly were Morgan's riflemen valued that Washington had personally assigned them to the Northern Department. At Saratoga, they would vindicate the faith he had placed in them. This division of the American army was commanded by General Arnold, a fire-eater on the field and a daring and skilled tactician. The American artillery train consisted of twenty-two pieces, enough for the business at hand though fewer than Burgoyne had.

On the British side, as Burgoyne's fortunes had ebbed, most of his Indian auxiliaries had melted back to Canada. A few Indians and Tory rangers remained, but they were now hounded by superior numbers of American scouts. British and German stragglers, and anyone wandering beyond the pickets, even briefly, was soon felled by American snipers. Except for occasional messages brought in by disguised couriers, Burgoyne was now cut off from the outside world and advancing blindly.

On the morning of the First Battle of Saratoga, the shortage of scouts was aggravated by a heavy fog which enshrouded both the British and American camps. Since crossing the Hudson, Burgoyne's army had moved in three parallel columns. He decided to continue advancing in this fashion, seeking to reach clear ground beyond the forest from where he would have a full view of the American lines that had thus far been invisible. He then planned to seize an unoccupied hill on the Americans' left flank and enfilade their position with his artillery.

The right column of his army, commanded by Brigadier-General Simon Fraser, numbered some 2,500 men. It included all the flank companies of the army – the plodding grenadiers and the mobile light infantry – together with the men of the 24th Regiment and a mixed body of Indians and Canadian and Tory regulars. This column advanced west for three miles from British headquarters at the Sword House and then swerved to the south. The centre column, which Burgoyne accompanied, was commanded by Brigadier-General Hamilton. It consisted of 1,400 men, all red-coated British regulars of the 9th, 20th, 21st and 62nd regiments. This column followed Fraser's line of march for a brief time and then turned in a south-easterly direction at a fork leading to the Great Ravine. After crossing the ravine, Burgoyne's column turned westward, maintaining this direction until contact was made with the Americans.

The left column was commanded by Burgoyne's senior German officer, the able, conscientious Baron von Riedesel. It was perhaps 1,800 strong and consisted mainly of the stolid, blue-coated Brunswick and Hanau troops – the regiments of von Riedesel, von Rhetz and von Specht and the men from Hesse-Hanau, as well as six companies of the British 47th Regiment. It advanced along the river road followed by the artillery train and the baggage.

For Burgoyne to expect concerted action between three blind columns operating in heavy forest over unfamiliar terrain was, to put it mildly, optimistic. According to his

On 19 September Burgoyne's army advanced in three divisions to attack the Americans who were entrenched on Bemis Heights, blocking the route to Albany. Fighting broke out in the clearing at Freeman's Farm when the British centre ran into Morgan's and Dearborn's marksmen. Although these riflemen were later substantially reinforced, Gates, fearing a non-existent threat to his right, still kept a large part of his army (some 4,000 men) idling behind their fortifications. The map shows von Riedesel's column after it had left the river road to go to Burgoyne's aid.

BRITISH
Main columns
1 Fraser
2 Hamilton
3 Von Riedesel

AMERICANS
Main columns
Fortifications
1 Morgan and Dearborn
2 Learned
3 Poor
4 Paterson
5 Glover
6 Nixon

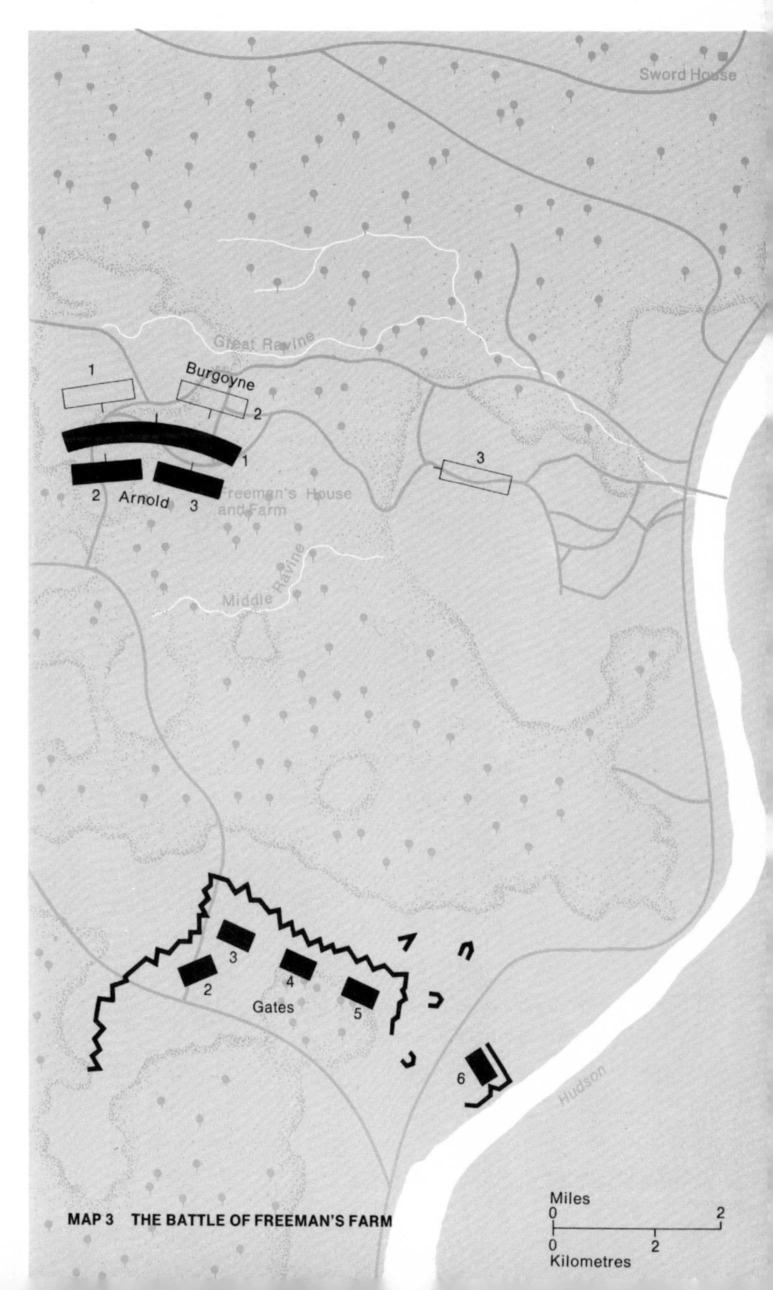

MAP 3 THE BATTLE OF FREEMAN'S FARM

battle plan the left column was to stay in readiness until signal guns were fired to indicate that Fraser's and Hamilton's columns were in position. Then a general advance was to begin in the direction of the American position (the size and shape of which were still unknown) until a suitable location near it could be found from which to launch a subsequent attack.

The American commander, Gates, whose scouts kept him advised of the latest British moves, still shied away from a pitched battle, no matter how tempting the odds. He preferred waiting behind his fortifications, hoping, perhaps, that Burgoyne would make a suicidal frontal assault such as had cost the British so heavily at Bunker Hill in the first real battle of the war. Arnold, supported by the more energetic members of the American officer corps, urged that the battle should be taken to the enemy, and the sooner the better. They believed in the ability of their men as all-round fighters, not just as snipers or defenders of fortifications, and they felt it was important to demonstrate their prowess in the battlefield.

Characteristically, Gates yielded – but only halfway. Rather than remain completely passive, or fully commit his army, he ordered Morgan's and Dearborn's marksmen to reconnoitre the enemy. Contact was achieved with the British centre at about 12.30 p.m. in a clearing of Freeman's Farm, by which the First Battle of Saratoga is also known.

After they had disposed of the British picket, Morgan's men advanced so hotly that they collided with Burgoyne's main body and were scattered in turn. Morgan, a tough leader with a formidable presence on the battlefield, rallied his men by means of his now-famous turkey-call whistle and was himself heartened by the arrival of two regiments of Continentals from Poor's brigade, despatched by Arnold. The rival forces glowered at each other from opposite sides of the clearing – Burgoyne's men on the north and Morgan's on the south. Burgoyne resolved to continue his advance, forming a line with the 21st on his right, the 62nd in the centre and the 20th on his left; the 9th were in reserve. The attack was met with a deadly fire from Morgan's riflemen, who inflicted heavy casualties on

the dense advancing line, and were close enough and skilled enough to pick off British officers.

It was a see-saw affair. As the British, hard-hit by Morgan's fire, began to fall back, the Americans charged across the field only to be driven off by the rallying British. The bloody clearing was the scene of a vigorous struggle for three hours, the British holding their own against Morgan's sharpshooters and increasing numbers of fresh men led into action by Arnold. Had not Baron von Riedesel taken it on his own initiative to rush to the aid of Burgoyne with part of his own division, the British centre column might well have been swamped. However, once von Riedesel arrived, American pressure slackened and Arnold's troops gradually disengaged, leaving Burgoyne, who had also been joined by Fraser, in possession of the field. At the height of the battle, Burgoyne's unaided centre column of 1,400 British infantry had held off as many as 3,000 American regulars – a tribute to the morale and fighting qualities that Burgoyne, whatever his failures as a strategist and tactician, was able to instil in the men serving under him.

Fearing a non-existent British threat to his right, the excessively timid Gates had kept another 4,000 combat-ready American troops out of action, idling behind his entrenchments while Burgoyne's columns advanced.

At this point the combination of dusk and the heavy forest terrain, as well as their own exhaustion, prevented the British from pursuing the retreating Americans.

In the British encampment, Burgoyne's men settled down to await further orders. At the rebel headquarters Arnold raged at the cowardice and slackness of Gates – and his anger was later heightened when Gates, in a moment of typical malice, omitted all mention of Arnold's role in the action in his report to the Continental Congress. Thus a bitter personal feud separated the American commander

above At the Battle of Bemis Heights, Burgoyne's flank rests in thick woods. To a traditional European foe, these woods would have represented a barrier; but to Arnold's men they were a convenient shield behind which to manoeuvre and snipe at the enemy. **right** A party of the local civilian volunteers and horses that were used to move the American artillery.

and his ablest subordinate just as the campaign approached its climax. On the British side, for nearly three weeks there would be no orders for further action against the enemy. Hunger, demoralization, and the bullets of unseen snipers were the only companions of Burgoyne's army during this uneasy interlude.

The Second Battle of Saratoga took place on 7 October when the 1,500 men of a British 'reconnaissance in force' advanced on the American left and were attacked, initially, by Poor and Morgan. Falling back under pressure from Arnold in the centre, the British then retired behind their fortifications; the Balcarres Redoubt held but Breymann's was overrun – which completed a disastrous day for Burgoyne and his men. The map shows the positions of both armies at the beginning of the battle.

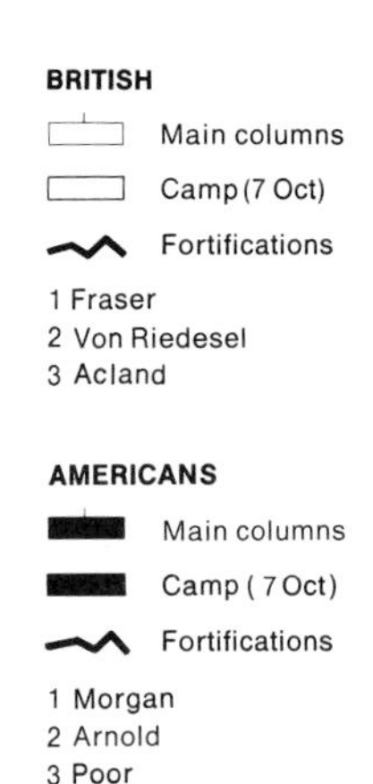

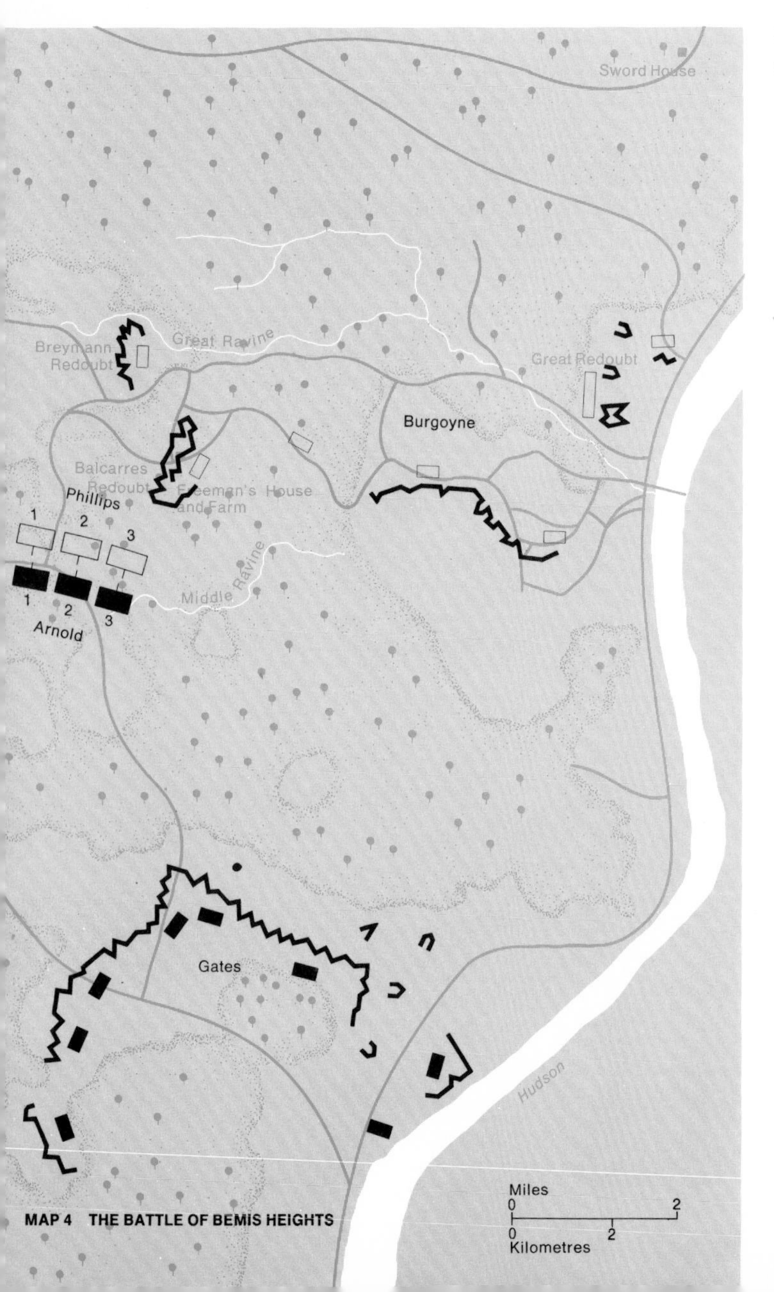

The Second Battle of Saratoga (Bemis Heights), 7 October 1777

In the wild hope that Major-General Sir Henry Clinton would send a rescue expedition up the Hudson, Burgoyne dug in and established his own fortified line. On its western extremity, and isolated from the main entrenchments, he placed the German grenadier battalion under Lieutenant-Colonel Breymann. This unit, badly shaken during the unsuccessful march to relieve Baum at Bennington, was thus placed in a vulnerable position: if it fell, the Americans would be able to force Burgoyne's right flank. Next to the German battalion, in the Freeman's Farm clearing where the first battle had been fought, he erected the Balcarres Redoubt, named after its commander, Lord Balcarres. Another stretch of fortified line covered Burgoyne's headquarters to Balcarres's right, and a 'Great Redoubt' to the north guarded his left flank and overlooked the Hudson.

On 6 October, Sir Henry Clinton did manage a light thrust up the Hudson from New York City. But Burgoyne was unaware of it at the time and was in no position to take advantage of it. In any event such a move was too little and too late. Its only effect was to jar Gates's already shaky resolution so that, when the time came to negotiate a surrender, he made some truly remarkable concessions to a powerless foe.

By 7 October Gates had been reinforced by another 4,000 militia, giving him a powerful advantage over Burgoyne's dwindling force, which was now also feeling a serious pinch in supplies and munitions. To remain stationary would merely ensure defeat for the British. Only a difficult retreat or a near-miraculous breakthrough of the American lines could salvage the situation. Burgoyne, still gambling, chose the latter course, but proceeded in such a way that scarcely any good could possibly come from it.

He ordered a 'reconnaissance in force' to be carried out – meaning, in effect, too many men for scouting and not enough for a serious engagement with the enemy – to test the ground around the American fortified line, which was still unfamiliar to him. This was to prepare the way for a full-scale attack on the American left, should all go well. It was the act of a man clutching at straws, and was made against the advice of his two best lieutenants, Fraser and von Riedesel, who urged an immediate retreat while there was still a chance of saving the army.

Fifteen hundred picked men were assigned to this most forlorn of tasks, led by Major-General Phillips, with Fraser and von Riedesel and accompanied by six 6-pounders, two 12-pounders and two howitzers. The column marched out of the British camp at noon and made for the American left. Gates had anticipated another desperate lunge by Burgoyne and deserves credit for keeping his men in readiness for it. But, again, all praise for leadership in the field must go to Morgan and Arnold. The latter, having lost his command after another violent squabble with

Gates, had been sulking in his tent until the guns began to blaze again. The intoxication of battle was the one thing Benedict Arnold could never resist, and so, though nominally without command, he mounted his horse, dashed to the sound of the guns, and ended up – through sheer force of personality – directing the action of 7 October.

The British force was deployed in a wheatfield, its flanks resting against thick woods. To a traditional European foe, these woods would have represented a barrier – to Arnold's men, they were a convenient shield behind which to manoeuvre, snipe at the enemy, and rush from at an opportune moment. Morgan's riflemen advanced to attack the British through the woods on the west while Enoch Poor's brigade attacked it from the east. Poor's men made the first contact.

The British right consisted of the light-infantry companies and the 24th Regiment, led by Fraser. In the centre stood von Riedesel with a picked contingent of his Germans, two British-manned 12-pounders, and the six 6-pounders of the Hesse-Hanau artillery company. Major Acland, commanding the British grenadier battalion, held the left.

The American attack began at approximately 2.30 p.m. Poor, with superior numbers, hit hard at the British grenadiers and crumpled them. Morgan, striking the light infantry on the right, also drove his opponents back. However, the indefatigable Fraser, mounted on a grey horse, dashed among his men and rallied them with a mixture of threats and persuasion. He cut a striking figure of courage in command – too striking, in fact, for he was soon noticed by Benedict Arnold.

One of Fraser's aides warned him that he was obviously a marked man, but was turned away with the remark, 'My duty forbids me to fly from danger'. A moment later, Fraser received a bullet through the body. He was carried from the field, mortally wounded. Deprived of their leader, his light infantry fell back.

While the British flanks reeled back in some disorder, Benedict Arnold dashed against the centre, manned by von Riedesel's blue-coated Germans. After repulsing the first attack, von Riedesel's men, exposed on both flanks, also fell back. Two thousand fresh New York militiamen now hurled themselves into the fray on the American side and the action in the wheatfield was decided. The outnumbered British column, badly mauled, sought refuge behind its breastworks. Burgoyne's last gamble, in less than an hour of combat, had cost him eight cannon and 400 officers and men killed, wounded or captured.

Very likely, if the cautious Gates had been in full control, the fighting would not have been as spirited. Almost certainly, it would have ended with the British withdrawal. But Gates hovered about his headquarters, out of harm's way, while in the field Arnold and the rebel troops were still eager to fight, and there were several hours of daylight remaining. Arnold and his men first charged headlong at the Balcarres Redoubt, but were beaten back.

'Well then,' Arnold exclaimed to a nearby officer, 'let us attack the Hessian lines.' By this he meant Breymann's Redoubt on Burgoyne's extreme right.

The thin line of Canadian irregulars covering the ground between Breymann's Redoubt and the rest of the British line had fled at the outset. Cut off, Breymann's men had panicked. Breymann himself may have been shot by one of them in the rout which ensued. By abandoning the redoubt, the Germans had exposed Burgoyne's right flank. The engagement had ended in disaster for the British.

Aftermath and Conclusion

Under cover of night, Burgoyne withdrew his army to a new position on high ground to the north. Fraser, who had asked that his body be buried in the Great Redoubt, died before dawn. He was interred on the evening of the 8th, attended by Burgoyne and the other surviving senior officers. On the morning of 9 October, cut off from any hope of reaching safety in Canada, the battered British army fell back to the heights of Saratoga.

Burgoyne had suffered approximately 1,000 casualties during the three weeks of fighting; the Americans, less than half that number. Within a few days, the rebel strength increased to almost 14,000 men, and Burgoyne, now outnumbered by at least three-to-one, was completely encircled and subjected to a lethal artillery bombardment. Gates, perhaps half disbelieving the astounding success of the Americans under his command (and possibly still casting a nervous eye in the direction of Sir Henry Clinton), quickly offered generous terms when Burgoyne refused to surrender unconditionally.

Although the British army was at his mercy, Gates made extravagant concessions – so extravagant that they were later repudiated by the Continental Congress. Under the terms of the surrender (which was diplomatically termed a 'Convention') Burgoyne's army was permitted to march out of camp with the full honours of war – as well it deserved after its gallant fight against hopeless odds. But Gates also agreed that the entire force would be returned to England on the condition that it would not serve again in America for the duration of the war. Since this did not preclude its relieving other British forces on duty elsewhere in the world, which could then be shipped to America, the Congress balked at this provision (understandably, if not very honourably) and many of the courageous men of the 'Convention Army' spent the rest of the war as abject prisoners, some never to return home. John Burgoyne, finished as a soldier, remained an ornament of London society in his later years, and continued to dabble in literature and politics.

Saratoga, wrote the nineteenth-century British historian, Sir Edward Creasy, 'insured the independence of the United States, and the formation of that transatlantic power which not only America, but both Europe and Asia, now see and feel.'

David Chandler at

AUSTERLITZ

Napoleon's victory against General Kutusov's Austro-Russian army reached a devastating conclusion when French cannon fire smashed into the frozen Moravian lakes and caused 2,000 fleeing Russians with their cannon and horse teams to sink abruptly from view. The brilliant campaign of 1805 ended at Austerlitz in great triumph for the First French Empire and its dynamic creator. Napoleon's martial genius had been displayed, and Continental Europe began to recognize its master.

1805

French artillery on Napoleon's left flank moving into position near the Santon Hill.

The Background to the Battle

The first long phase of the wars deriving from the French Revolution came to an end with the Peaces of Loeben and Amiens in 1801 and 1802 respectively. For France and Great Britain, however, the ensuing period of tranquillity proved hardly more than a breathing space, and in May 1803 war again broke out. At the resumption of hostilities the bulk of the French army was stationed along the English Channel coast preparing to invade England, but in September 1805 Napoleon gave orders for the Camp of Boulogne to be broken up, and the 'Army of England', rechristened 'the Grand Army', was soon thereafter deployed along the Rhine. Indications of a burgeoning coalition hostile to Napoleon had already taken concrete shape. Both Austria and Russia had had slights and humiliations to avenge, and persuasive diplomacy by the British Prime Minister, William Pitt – aided by the transfer of no little British gold – had rapidly transformed these aggressive tendencies into the Third Coalition, formed in April 1805.

Typically, Napoleon determined to get his blow in first. By late September he had massed 210,000 troops between Mainz and Strasbourg. Learning that the main Austrian army was being collected in north Italy (as in 1796 and 1800), and that 72,000 Austrians under Archduke Ferdinand, a son of the Emperor, and General Mack, were advancing on Ulm without awaiting the arrival of promised Russian aid, the Emperor launched a telling blow. Luring Mack deeper into the Black Forest defiles by means of provocative cavalry demonstrations (operating from Strasbourg under the French Marshal Prince Murat), Napoleon launched the rest of his army into a great wheeling march from the Rhine to the Danube on 25–26 September. The French army swept through Germany on a broad front and at a speed that the Austrians generals could barely comprehend. By 6 October the reconcentrating French forces were crossing the Danube, severing Mack's links with both the Russians and Vienna, and isolating his army. Next, centring their operations around Augsburg, the French turned west. The result was that some 60,000 Austrians were encircled at Ulm and a number of outlying centres, and surrendered: it was almost the perfect 'bloodless victory'.

It remained to mete out similar treatment to the Russians commanded by General Kutusov, who were currently near the River Inn. The Russians proved harder to catch than Mack, however, and their withdrawal eastwards, and then to the north of the Danube, increasingly frustrated Napoleon and drew him further and further from his bases. Even the capture of Vienna on 12 November proved worthless strategically, for Kutusov still refused to fight, eluding several French traps (bungled though two of them were by Murat). The sole Russian aim was to unite with General Buxhowden and the Tsar near Olmütz before turning on their pursuers. After fighting rearguard actions at Hollabrunn and Schongrabern, Kutusov at last joined his

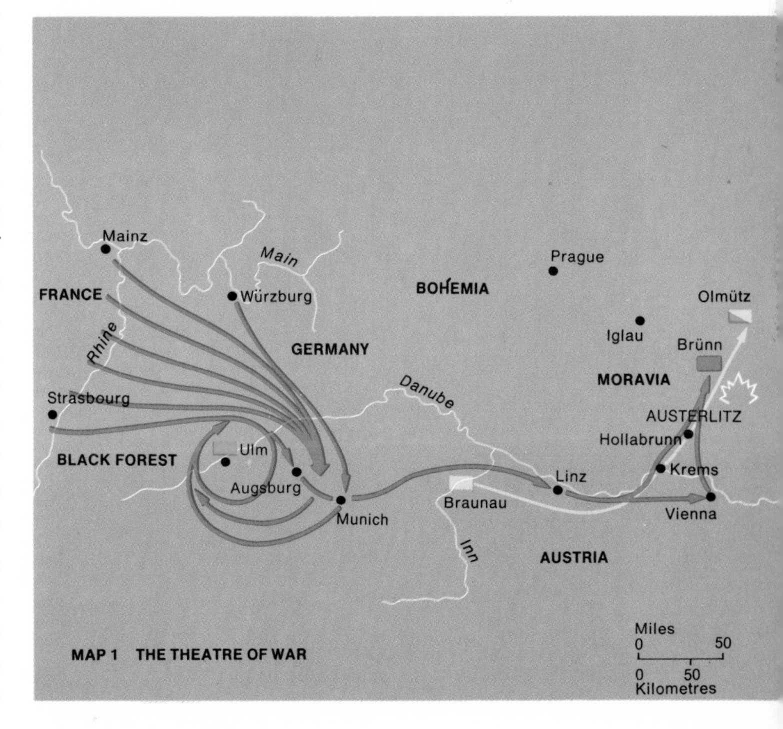

Napoleon began his campaign of 1805 against the forces of the Third Coalition by securing the bloodless capitulation at Ulm and various outlying centres of some 60,000 Austrians under General Mack, who found themselves suddenly encircled and cut off from Vienna and their Russian allies. Next, Napoleon pursued General Kutusov's Russian army but could not prevent it from crossing the Danube and linking up with General Buxhowden and the Tsar near Olmütz.

Main French advance (Napoleon)
Austrians (Mack)
Russians (Kutusov)
Allied Austro-Russians

colleagues on 20 November. The scene was now almost set for the great 'Battle of the Three Emperors'.

All in all Napoleon's strategic position in late November was far from favourable. Since his *blitzkrieg* success against General Mack, the pursuit of the elusive Kutusov had placed a great strain on the Grand Army. No less than 450 miles now lay between the French spearhead and the bases along the Rhine, and the wastage caused by incessant marching and the need to garrison men along the line of advance had greatly reduced Napoleon's battle power. The original 210,000 had shrunk to 55,000 under the Emperor's immediate command. Furthermore, the threats to the French position were multiplying. The reinforced Austro-Russian army facing them now numbered 85,000 men; from the south, the Archdukes Charles and John of Austria, sons of the Emperor Francis, were closing in with 122,000 more; and, perhaps worst of all, there were ominous signs that Prussia was about tc join the Allies – an event which could mean another 200,000 opponents to the north. This left Napoleon with a difficult decision. To continue the advance, or to remain halted, were equally dangerous propositions; on the other hand, to retreat would be to court catastrophe.

Napoleon, however, was often at his best in a crisis. If he was going to fight for his very survival, it should at least be on ground, and at a time, of his choosing. With great finesse he dangled the bait of an easy victory before his opponents. Feigning weakness, French cavalry fled on contact with Cossack patrols near Olmütz. The Army then proceeded to evacuate the town of Austerlitz and, even more pointedly, abandoned the key Pratzen Heights, showing every sign of disorder. When the Allies tentatively offered an armistice (to give the Archduke Charles time to appear), Napoleon went out of his way to be courteous to the bombastic Russian envoy, Count Dolgorouki, even escorting him personally to the outposts. These associated deceptions changed Allied caution to overconfidence, and on 1 December their massed columns marched west from Olmütz and occupied the Pratzen.

By this time Napoleon's countermeasures were nearing

above An officer of Napoleon's Guard Lancers. **right** An elegant French cannoneer of the period. Following the reform of the French artillery from 1774, field guns were reduced to four types, 12-, 8- and 6- (or 4-) pounders, and 6-inch howitzers. Gun carriages and other equipment were also standardized and made partly interchangeable.

completion. Hidden behind a screen of cavalry, Marshals Bernadotte and Davout were already hastening from Iglau and Vienna respectively with their I and III Corps. The bulk of the former unit arrived on 30 November, and by dint of Herculean marching Davout's leading division was within range late on the 1st. This redrew the odds at almost eight to seven in terms of soldiers, but the Allies retained a convincing two-to-one superiority in artillery, and were marginally better supplied than the French.

The Rival Armies

The comparative strengths of the rival armies at Austerlitz were approximately as follows:

French		**Allies**	
Infantry and artillerymen	64,500	Infantry and artillerymen	76,400
Cavalry	8,700	Cavalry	9,000
Total		*Total*	
73,200 men and 139 guns		85,400 men and 278 guns	

above A colonel-general of the French hussars, much noted for their dash and swagger. **left** 'Serrez les rangs!' A sergeant of Napoleon's army orders his men to close ranks during the fight. The Grand Army of 1805 was originally organized into seven corps (including one of Bavarians) of which four, together with the Cavalry Reserve and the Guard, fought at Austerlitz. Each corps was a miniature army, with its own staff, infantry, cavalry and artillery components, trains and ambulance. In size and composition they varied greatly but it was Napoleon's intention that a corps should be able to defend itself against any number of opponents for up to a day's fighting.

General Kutusov

Austro-Russian Commander-in-chief

Mikhail Hilarionovich Golenischev-Kutusov was titular commander-in-chief of the Austro-Russian forces at Austerlitz, although the presence of both the Tsar Alexander I and the Emperor Francis I led to considerable intrusions upon his authority.

Aged sixty in 1805, Kutusov had entered the Russian army when he was sixteen, after studying at Strasbourg. Serving in the artillery of Catherine the Great, he took part in campaigns in Poland, Turkey and the Crimea, where he lost his right eye. In 1790 he played a key role in the capture of Ismail under the command of the great Suvorov. Promoted to lieutenant-general the following year, he was sent as ambassador to Constantinople in 1793. Later years found him commander of the Ukraine, and then military governor of St Petersburg. In 1805 he was selected to lead a Russian army to the assistance of Austria. Physically corpulent and clumsy, by nature he was well-meaning but also wary and astute, preferring caution to rashness. His bravery and popularity with the rank-and-file were well known, but his reputation as a battle-commander suffered after Austerlitz and Borodino (1812). His handling of the pursuit of the French from Moscow also earned much criticism, but was on the whole well advised. Clausewitz declared that 'he could flatter the self-esteem of both populace and army'. Count Philippe de Ségur opined that Kutusov's 'valour was incontestable, but he was charged with regulating its vehemence according to his private interests. His genius was slow, vindictive and above all crafty . . . knowing the art of preparing an implacable war with a fawning, supple and patient policy.' Such was Napoleon's principal opponent in both 1805 and later at Borodino in 1812.

below The Allied outflanking column crosses the Goldbach. The aim of this powerful force was to block the Vienna road to the south and then sweep northwards, rolling up the French line.

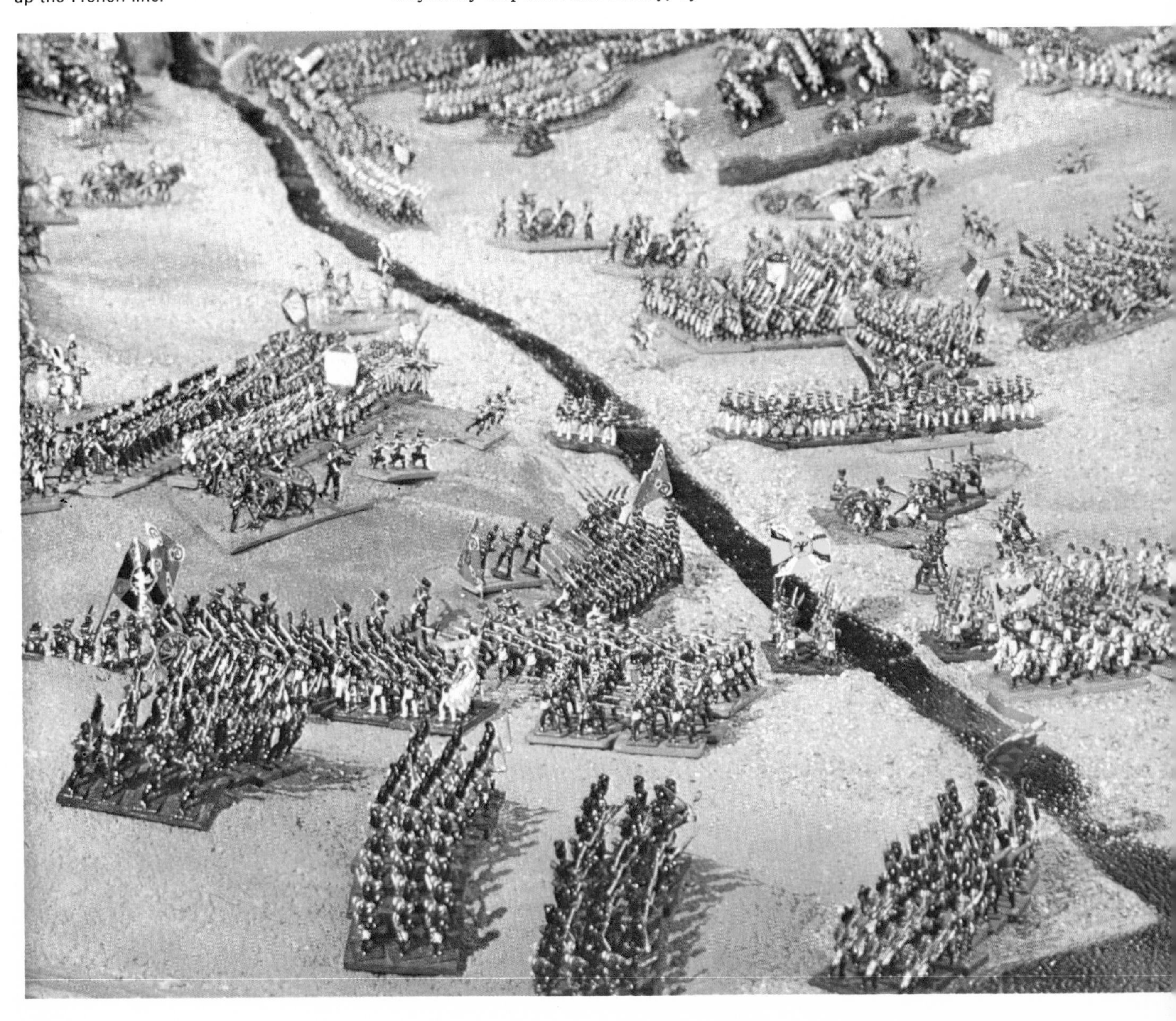

The Emperor Napoleon I

French Commander-in-chief

On 2 December 1805, Napoleon had been crowned Emperor for exactly one year. He was thirty-six years old and at the height of his powers.

As a young officer, he had seized the opportunity afforded by the French Revolution to emerge from the obscurity of his Corsican background and his early years as an impecunious lieutenant of artillery in the Bourbon Royal Army. In 1799 he led the *coup d'état de Brumaire* against the Directory, and emerged as one of three Consuls. Soon he was undisputed First Consul, and in 1800 he accompanied the Army of Reserve over the Alps to fight at Marengo against the Austrians.

The brief years of comparative peace that followed saw the start of the great series of legal and institutional reforms which were Napoleon's greatest constructive achievement. In recognition of his endeavours and in the hope of establishing a stable succession, a national plebiscite voted overwhelmingly in favour of his elevation to the throne, and on 2 December 1804 he was duly crowned Emperor in the Cathedral of Notre-Dame in the presence of Pope Pius VII.

By 1805, Napoleon had both the French people and their army wholly in thrall. His charisma was compelling – even sworn foes succumbed to his charm and genius. His frown was as dreaded as his approval was courted. 'So it is that I,' recalled one hard-bitten general, 'who fear neither God nor devil, tremble like a child at his approach.' Orders, titles and wealth were lavished on the faithful; political opponents were either won over or ruthlessly eliminated. But the First French Empire and its presiding genius had still to undergo the acid test of full-scale continental war against the old established monarchies of Europe.

below The armies were arrayed along a front almost five miles wide extending from the Santon Hill in the north to a group of frozen lakes in the south near Telnitz. To lure the Allies into attacking him in the south Napoleon placed a mere token force there under Le Grand (though the latter was promised support from Davout's men, still rapidly advancing from Vienna); the decisive French force was deployed around Puntowitz and the Zurlan Hill, ready to storm the Pratzen Heights once the Allied left and centre were committed in the south. The Allies were totally deceived, and four columns under the general command of Buxhowden (those of Kienmayer, Doctorov, Langeron and Przbysewski) made ready to cross the Goldbach at first light the next day.

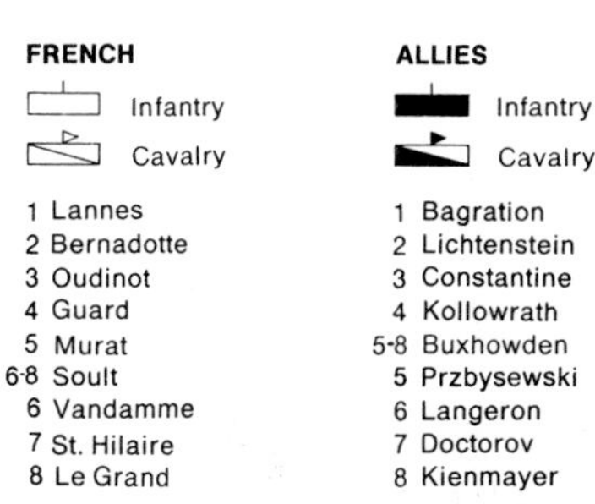

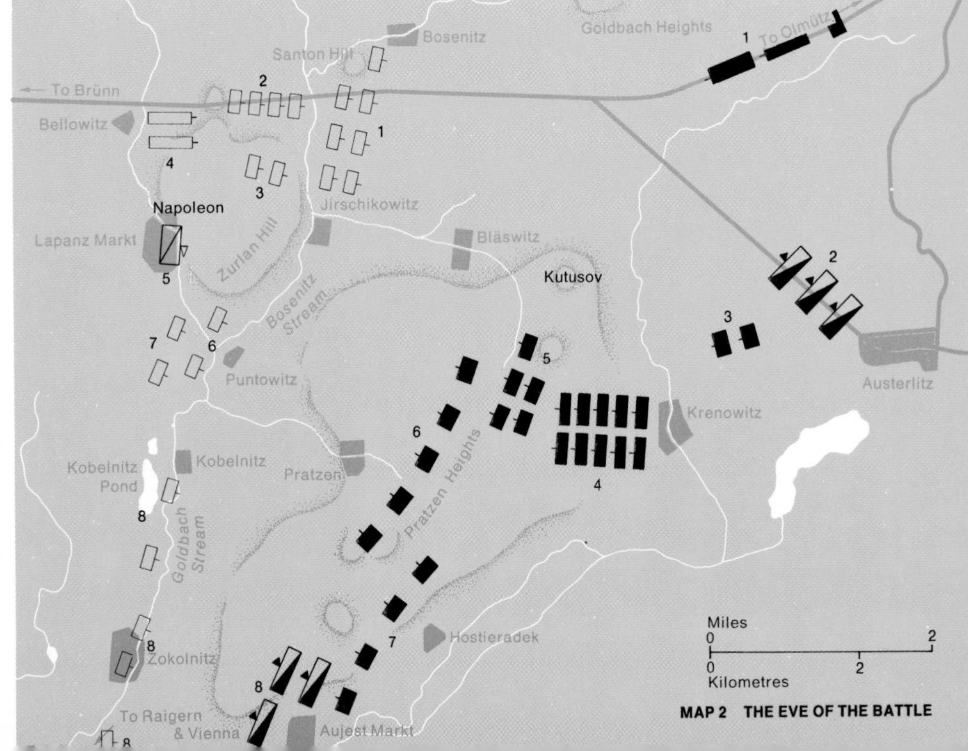

The field of Austerlitz extends from the Santon Hill (north of the Brünn–Olmütz high road) to the frozen lakes near Telnitz – a distance of almost five miles. (These Moravian towns are now in Czechoslovakia and appear respectively on modern maps as Slavkov – Austerlitz – Brno, Olomouc and Telnice. Throughout this account the old names are used.) Two streams, the Goldbach and the Bosenitz, form a confluence dividing the Pratzen Heights to the east from the French-held Zurlan Hill to the north-west. Around this hill Napoleon had concentrated all of 65,000 men, leaving only one and a half divisions of Soult's IV Corps to hold three miles of front running through the villages of Kobelnitz and Zokolnitz to Telnitz on the extreme right. Hidden behind the flank, however, around Gross Raigern, were the footsore infantry of Davout's newly arrived III Corps, which appeared after dusk on the 1st.

The French infantry at this time comprised three battalions, each of nine companies (one light, one of grenadiers, the remainder line). A company held 140 officers and men at full establishment. Some regiments – perhaps a quarter – were entirely made up of tirailleurs, or light infantry. The standard firearm for all formations was the 1777 Charleville musket, of .70 calibre, measuring fifty inches and capable of firing two or three rounds a minute. Each soldier carried twenty-four rounds in a cartridge pouch, and had a triangular-section bayonet and a short curved sword, besides a skin-covered pack and blanket roll. Officers carried swords and pistols. In action, the light infantry fought as skirmishers, the remainder in combinations of column and line.

The French cavalry was of three main types. There were the 'heavies', cuirassiers or carabiniers, who were armed with long straight swords, pistols (some with carbines) and wore breast- and back-plates and steel helmets with horse-hair plumes. Then there were the dragoons, armed with sabres and carbines, but who wore no armour except for their helmets; and the light cavalry, mainly hussars, dressed in a profusion of splendid uniforms and much noted for their dash and swagger. The latter were armed with sabres and pistols. There were also lancers and chasseurs.

The artillery was equipped with four types of gun – 12-, 8- and 6- (or 4-) pounders, and 6-inch howitzers. The 12-pounders were deployed at corps level or in the reserve, the 8-pounders were allocated to divisions, and the 6- or 4-pounders to individual regiments. Each cavalry division had two companies of horse artillery, and more were held in the artillery reserve.

The grand tactics of the Napoleonic battle were based on close co-operation between the various elements of the army, and the massing of superior force at a suitable point ready to break the enemy line. A preliminary bombardment would be unleashed against the enemy, under cover of which light infantry moved forward to open a sniping fire. On many occasions, an attack by cuirassiers was then unleashed to defeat the enemy's cavalry and force its infantry battalions to form square, so providing an ideal target for horse artillery accompanying the cavalry. Protected during these developments, the infantry columns then hastened forward to close range, formed into line or charged in with the bayonet, thus driving a wedge into the enemy's tiring front. Once a gap had materialized, the massed light cavalry and dragoons would come forward, sabres relentlessly rising and falling, to exploit the local breakthrough and convert the enemy's setback into a full-scale rout.

By the afternoon of 1 December, the bulk of the Allied army was massing on the Pratzen Heights and near Aujest Markt to the south, while a secondary force camped near the Olmütz-Brünn highway in the north, facing the Santon Hill. Headquarters and the Russian Imperial Guard were situated at Krenowitz, a short distance to the east. Approximately two-thirds of the forces present were Russian. Both Allied armies included regiments of grenadiers and jägers (light infantry) as well as line infantry, and habitually fought in linear formations. The cavalry comprised the same main types as the French, with the addition of uhlans (lancers) and detachments of semi-irregular Cossacks and Tartars, armed with lances, pistols, sabres and carbines. Many Russian artillery batteries held twelve cannon apiece. The Russian Guard Corps comprised the Preobzhenski, Semenovski and the Grenadiers, Guard

left Napoleon gives Marshal Soult the order to advance on the Pratzen.

Jägers, the mounted Garde du Corps, the noble-born Chevalier Guard, and detachments of élite light cavalry.

It was important for Napoleon to influence the form of the battle, which his skill had induced his enemies to undertake. His plan was to lure the Allies into an all-out attack against his weak right flank in the direction of the highroad to Vienna, which appeared to constitute the sole line of retreat. In fact Napoleon had already designated a safer route running west through Brünn, but this switch, like the arrival of reinforcements, was carefully concealed from the Allied patrols.

To lure the foe into attacking in the south, General Le Grand was given only a skeleton force to hold the line of the Goldbach; his orders were to withdraw northwards if necessary, but he was assured that he could count on aid from Friant's division of Davout's III Corps, newly arrived in the rear of the right wing. Secondly, Napoleon entrusted Marshal Lannes's V Corps, supported by Prince Murat's Cavalry Reserve and part of Bernadotte's command, with the task of holding the area of the Santon on the left, which had been fortified. Like Le Grand's, this was basically a defensive role.

Meanwhile, in the vicinity of the Zurlan Hill, the Bosenitz stream and the neighbouring village of Puntowitz, Napoleon deployed his decisive force – two divisions of Marshal Soult's IV Corps (commanded by Generals St Hilaire and Vandamme) supported by the Imperial Guard, Oudinot's Grenadier Division, the reserve artillery, and parts of Murat's cavalry and Bernadotte's I Corps. The plan was to wait until the Allies had committed all their left wing and centre against the French right, and then to unleash Soult against the Pratzen. Once the plateau was firmly in French hands, the reserves would sweep forward into the gap in the Allied line to envelop either the sundered right or left of the Allied army.

As the Russian and Austrian generals gathered at Allied headquarters to consider their final plans in the early hours of 2 December 1805, the prevalent atmosphere was one of anticipation. Their prospects of victory seemed assured. They believed that Napoleon's willingness to consider an armistice, and the French surrender of the Pratzen, proved that the Emperor's nerve had cracked. In the conference that followed, only one cautious voice was raised. General Kutusov suggested that appearances might prove deceptive. The hotheads around him, however, paid little heed to their titular commander-in-chief. The Austrian General Weyrother and the Tsar's aides knew that Alexander was determined to fight at once, and as there were three Russians to every Austrian in the Allied army the known hesitations of the Emperor Francis could also be ignored. Kutusov was not prepared to oppose the consensus, and as Weyrother droned out the interminable orders, General Langeron noted that the old warrior, 'half-asleep when he arrived, at length fell into a sound nap before our departure'.

The Allied plan of attack called for the division of the army into three equal parts. All of 45,000 men were to gather on the left in four columns (commanded by Generals Kienmayer, Doctorov, Langeron and Przbysewski) under the overall direction of General Buxhowden. This large array was to cross the Goldbach, block the Vienna road, and storm the French-held villages; this done, it was to

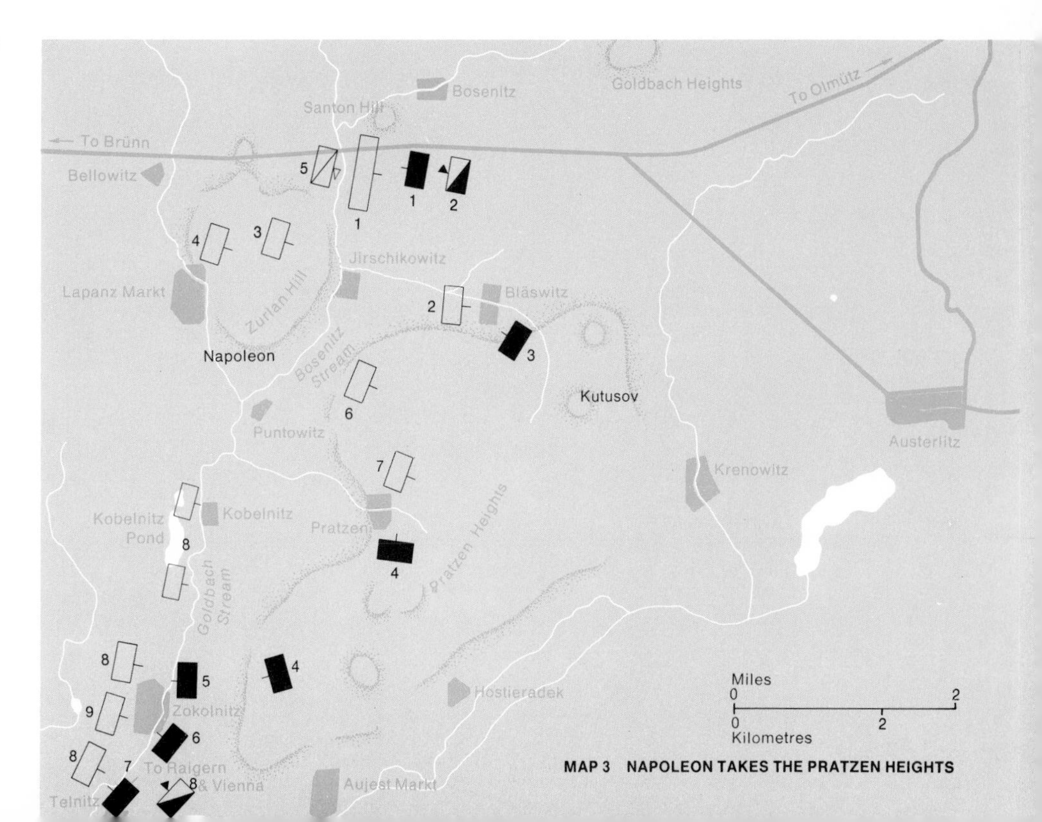

This simplified map shows the direction of Soult's surprise thrust through the Allied centre. As soon as Buxhowden's columns were fully engaged in the south at Zokolnitz and Telnitz, and Kollowrath had moved down to join them, Napoleon unleashed two of Soult's divisions, commanded by St Hilaire and Vandamme, against the Pratzen Heights. Although Kollowrath sent back troops to challenge the French assault, this move by Napoleon effectively bisected the Allied line. Further north Bernadotte was ordered to attack Bläswitz while Lannes and Murat fought a gallant holding action near the Santon Hill against the superior numbers of Bagration and Lichtenstein.

FRENCH	ALLIES
Infantry	Infantry
Cavalry	Cavalry
1 Lannes	1 Bagration
2 Bernadotte	2 Lichtenstein
3 Oudinot	3 Constantine
4 Guard	4 Kollowrath
5 Murat	5-8 Buxhowden
6-8 Soult	5 Przbysewski
6 Vandamme	6 Langeron
7 St. Hilaire	7 Doctorov
8 Le Grand	8 Kienmayer
9 Friant	

right The cavalry battle begins near Bläswitz. **below** Murat's cavalry massing on the Zurlan Hill before joining Lannes' troops against the combined forces of Bagration and Lichtenstein's cavalry squadrons. **far right** The Allied left begins to disintegrate.

sweep remorselessly northwards, rolling up the French line, while a fifth column (15,000 men under General Kollowrath) descended from the Pratzen to seize Puntowitz and thus knock the hinge out of the buckling French line. In the meantime, a second force, 17,600 strong under General Bagration and Prince Lichtenstein, would close up to the Santon and keep the French left in play. The Russian Imperial Guard (8,500 élite troops under Prince Constantine) would be held back to form a central reserve.

As the 1 a.m. conference of dozing or over-confident Allied generals proceeded, across the mist-shrouded valley the French were also engaged in martial preparations. Earlier in the night, Napoleon had walked through his ragged and hungry troops' bivouacs to gauge their morale. He was more than reassured. The men had cheered their Emperor as he strode past their camp fires, and had formed a torchlight procession to escort him back to his rough quarters. 'This has been the finest evening of my life,' he was heard to murmur. He was also confident – from the reports of his patrols – that the foe was about to comply with his wishes almost to the last letter. His deadly traps were set and they were marching to their doom, though few apart from Napoleon yet suspected it. 'Before tomorrow evening this army will be mine,' he had predicted late on 1 December.

The Course of the Battle

Both armies were astir by 4 a.m. The Allies found the dense fog a grave inconvenience as they formed up, but by seven o'clock Buxhowden was attacking Telnitz and Zokolnitz, and eventually became their master. This coincided with the arrival of General Friant's footsore but battle-ready division from Gross Raigern, and Le Grand was able to regain a little lost ground. This in turn induced Buxhowden to summon Kollowrath ahead of schedule, and so by 8.30 a.m. the last Allied battalions were quitting the Pratzen and moving south. The mist was slowly clearing, but it still clung to the valley floor.

These developments did not go unobserved by Napoleon on the Zurlan, also clear of mist. Beside him waited an anxious Berthier, Chief of Staff, and Marshal Soult, whose two divisions, liberally fortified by a triple brandy issue, waited in concealment near Puntowitz in the valley below.

Timing was of the very essence. The enemy must be afforded sufficient time to weaken his centre. At 9 a.m. the moment was deemed to be right. The advance was ordered.

The drums beat the *pas de charge*, and the bright morning sun glinted menacingly from the serried ranks of bayonets as Vandamme and St Hilaire emerged from the fog-filled valley.

The Allies were taken completely by surprise. Kutusov urgently recalled part of Kollowrath's column, but it returned too late to prevent the French from occupying the Pratzen's summit. Napoleon now ordered Bernadotte to occupy Bläswitz in support of Soult's initial success.

By this time a new conflict was flaring up around the Santon as Bagration closed with Lannes. The defences stood the test, but it proved advisable to sustain V Corps' raw conscripts with the cavalry. The outcome was a fierce cavalry action in which the French cuirassiers decimated Lichtenstein's massed squadrons. Meanwhile, the main battle raged on away to the south with varying fortune, Napoleon reinforcing his tiring right wing with Oudinot's grenadiers.

Back in the centre, the crisis was at hand. By 10.30 a.m. IV Corps was under heavy attack from three directions as Kutusov hounded up more forces recalled from his left. However, Marshal Soult was at hand to site six 12-pounder cannon which redressed the adverse balance for the time being. The French remained in over-all control of the Pratzen, and Napoleon moved the Guard and his headquarters nearer to the fighting.

All was not yet over in the centre, however. About 1 p.m. the weary French were suddenly attacked by the Russian Imperial Guard, which, resplendent in brass-faced mitre caps and green, red and white uniforms, came sweeping up the last 300 yards to the crest at the full charge, accompanied by Guard cavalry. The first French line gave way and two battalions (one having lost its eagle-colour) broke and fled, almost trampling the Emperor and his staff on their way to the rear. Fortunately the Guard Cavalry was at hand. Forward went Bessières, followed by General Rapp (Napoleon's favourite aide) at the head of a second wave. The Russians were soon put to flight, and Bernadotte, seeing how matters stood, intelligently rushed up a division to sustain the near-exhausted French centre. The Russian Guard left behind 500 dead and over 200 prisoners, many of them from the white-coated Chevalier Guard, the Tsar's personal escort. 'Many fine ladies of St Petersburg will rue this day's work,' Napoleon grimly commented.

The crisis in the centre being averted, the day was virtually won. It still remained, however, to convert victory into triumph. Napoleon now sensed that the opportunity lay to the south, and ordered Soult's survivors and the Guard to envelop Buxhowden's isolated masses from the eastern slopes of the Pratzen, while Davout and Le Grand made one more frontal attack in the valley to pin the foe. 'Let not one escape!' was Marshal Davout's grim order. The time was now 2.30 p.m.

In another hour it was almost over. Przbysewski's and half of Langeron's men were forced to surrender.

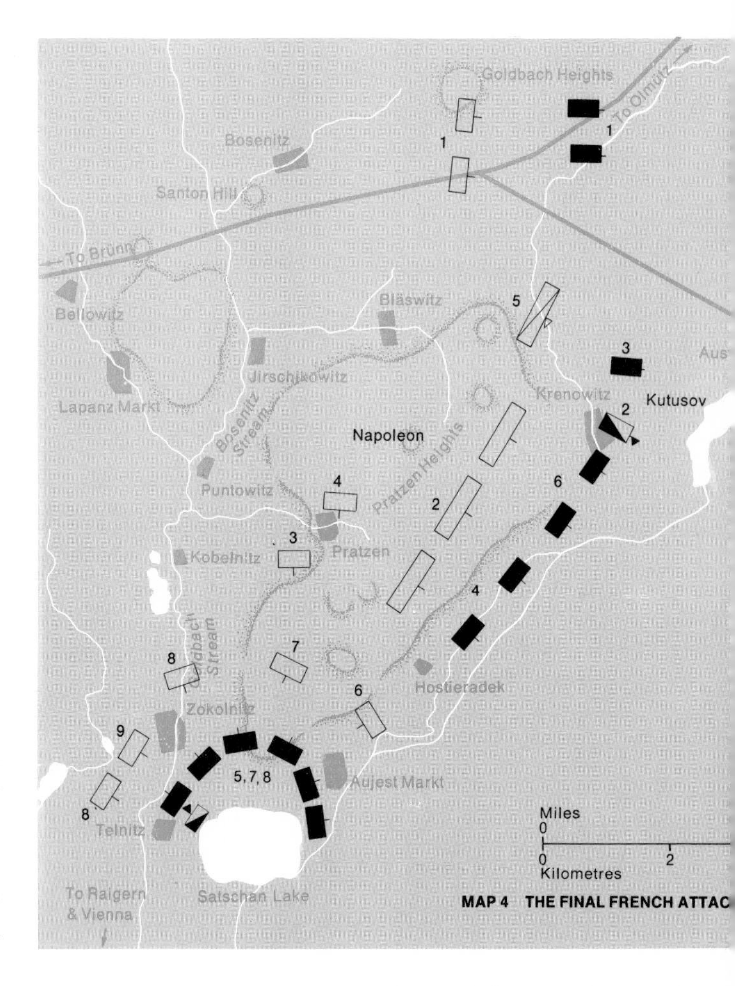

Napoleon delivered his final crushing attack in the south, where Buxhowden's force was enveloped and largely destroyed. In the north Bagration was soon in full flight and in the centre the Allies evacuated Krenowitz and Austerlitz as part of a general withdrawal. The battle was over; at a cost of just over 8,000 men the French had killed and wounded 16,000 Allies and taken 11,000 prisoners.

Buxhowden and part of one column slipped out of the trap, but for Doctorov there was no hope. Penned against the frozen lakes near Telnitz he ordered his men to scatter over the ice in search of safety. This was seen by Napoleon. At his order, a battery of artillery was rushed to the fore of the overlooking Heights, and its salvoes smashed into the ice. Perhaps 2,000 Russians were drowned, and thirty-eight Russian cannon and their horse-teams sank like lead to the bottom of the lakes. The battle was over.

Bagration was by then in headlong retreat from the Santon; Tsar Alexander, the Austrian Emperor, General Kutusov and their suites, escorted by the Russian Guard, evacuated Krenowitz and Austerlitz; the rest of the army had ceased to exist as a viable concern. The last guns fell silent at four o'clock, as a snow blizzard spread a merciful cloak over the stricken field. Time could at last be spared for the many, many wounded. The French had lost approximately twelve per cent of the Grand Army: 1,300 killed, 6,490 wounded and 500 missing. They had killed and wounded 16,000 of the Allies, and taken 11,000 prisoners – some thirty-three per cent of the whole.

Aftermath and Conclusion

Napoleon had gained his decisive victory, snatching a great success from the jaws of strategic defeat. He had inflicted three times the casualties his army had sustained, and taken 180 guns and forty-five colours. The next day a humbled Emperor Francis came to beg for an armistice and peace negotiations, which later in the month resulted in the Peace of Pressburg. As for the Tsar, his sole idea was to retire towards Poland. 'I am going away,' he wrote to his conqueror, '. . . yesterday your army performed marvels.'

Napoleon had good reason to be satisfied. 'Soldiers, I am pleased with you,' began the victory bulletin. Lavish rewards were later distributed: between them the marshals and generals received two million golden francs; pensions were provided for the widows of the fallen; orphans were to be formally adopted by the Emperor, and allowed to add Napoleon to their baptismal names; they were also to be educated at the state's expense.

At one stroke the Third Coalition had been destroyed. Prussian emissaries, who had presented themselves in Vienna in late November bearing an ultimatum from their master, King Frederick William IV, hastened now to offer profuse congratulations. Napoleon was not fooled, sardonically remarking that the letter appeared to have been recently readdressed. Prussia would hear more from the French in 1806. In far-off England, news of the disaster to all his hopes caused William Pitt to declare: 'Roll up the map of Europe.' Broken-hearted, he was dead within a few months.

The brilliant campaign of 1805 thus ended in a great triumph for the First French Empire and its dynamic creator. Napoleon's martial genius had been convincingly displayed, and Continental Europe began to recognize its master. Many trials and successes still lay ahead of the newly blooded Grand Army and its thirty-six-year-old commander, but none of its triumphs would surpass the achievements of 2 December 1805.

James Lawford at

WATERLOO

At Mont St Jean near the Belgian village of Waterloo, Napoleon Bonaparte wagered his Empire against the might of two Allied armies. On the eve of the battle Napoleon's rivals were camped some eight miles apart. But next day, before he could crush the Duke of Wellington's Anglo-Dutch force he was taken in the flank by Blücher's Prussians and his main army, originally 74,000 strong, was brutally dismembered.

1815

At a crucial phase in the battle – with Napoleon now desperate to smash Wellington's army before the Prussians' arrival – the Duke takes temporary refuge inside one of his infantry squares.

The Background to the Battle

On 31 March 1814 the Allied armies of the Tsar and the King of Prussia entered Paris, and Napoleon was forced to abdicate. He was exiled to Elba, but within less than a year he had returned. Having escaped from Elba he managed swiftly to regain France. The Allied Powers – Britain, Russia, Austria and Prussia – that had brought about Napoleon's downfall the previous year declared him an outlaw, and prepared themselves again for war.

By May 1815 two large Allied armies had formed in the Netherlands, a Prussian army under Marshal Blücher about 120,000 strong, and, under Wellington, a composite army of equivalent size with contingents from Britain, the Netherlands, Brunswick and Hanover. (For convenience the latter will be referred to as the Anglo-Dutch army, although George III's Hanoverian subjects played a considerable part in the fighting that followed.) The quality of the Anglo-Dutch army was by no means uniform; there were numbers of raw recruits and many of the Belgians in the regiments from the Netherlands cherished a warmer affection for Napoleon than for the Allies. Wellington had his headquarters at Brussels while Blücher had established his at Hannut, twenty miles west of Liège.

Napoleon, watching his enemies gather, resolved to strike first and to do so in the north. After detaching an unavoidable minimum of troops to watch his other frontiers, he had available a field army of about 120,000. Despite the odds against him, he hoped to snatch success by a lightning offensive. He proposed to invade the Netherlands, separate Wellington and Blücher, and trounce each in turn before they had time fully to concentrate their armies.

On 14 June he set his army in motion. The next day he crossed the frontier into the Netherlands, swept over the Sambre by Charleroi and marched on Brussels. The Allies knew nothing of his presence until, in the early hours of the morning of the 15th, he thrust aside their outposts on the border; the speed and audacity of the great master had caught them off balance.

As soon as he heard what had happened, Blücher ordered his army to concentrate near Ligny. Unfortunately, he despatched the fattest officer on the slowest horse in the Prussian army to tell Wellington what he intended, and the information was somewhat late in arriving. Wellington had already learned that the French were advancing and, uncertain of their aim, that evening had ordered his army to assemble in the area Nivelles–Braine le Comte–Mont St Jean. General Constant de Bebecque, principal staff officer to the Prince of Orange, realized that the crossroads at Quatre Bras would be a vital link with the Prussians and on his own initiative directed Perponcher's Belgo-Dutch division to defend this point.

On the morning of the 16th, Napoleon ordered a general advance on Brussels. To his joy, a large Prussian force had been identified near Ligny. He decided to crush it himself, while Marshal Ney with a wing of the army drove on

On 14 June 1815, Napoleon crossed the Sambre by Charleroi and marched on Brussels. His first aim was to separate the Allied armies of Wellington and Blücher and destroy each in turn. At Ligny on the 16th he defeated Blücher heavily and the Prussians withdrew to Wavre. To the west Ney attacked Wellington at Quatre Bras but on the next day the Anglo-Dutch army broke away north to Mont St Jean. Napoleon then sent Grouchy after Blücher and himself prepared to do battle with Wellington. The latter, assured of Blücher's help on the 18th, placed 15,000 men at Hal to guard the Mons–Brussels road and ordered his main army to bivouac in the Mont St Jean position. The map records the movements of the Allied and French armies up to the night of the 17th.

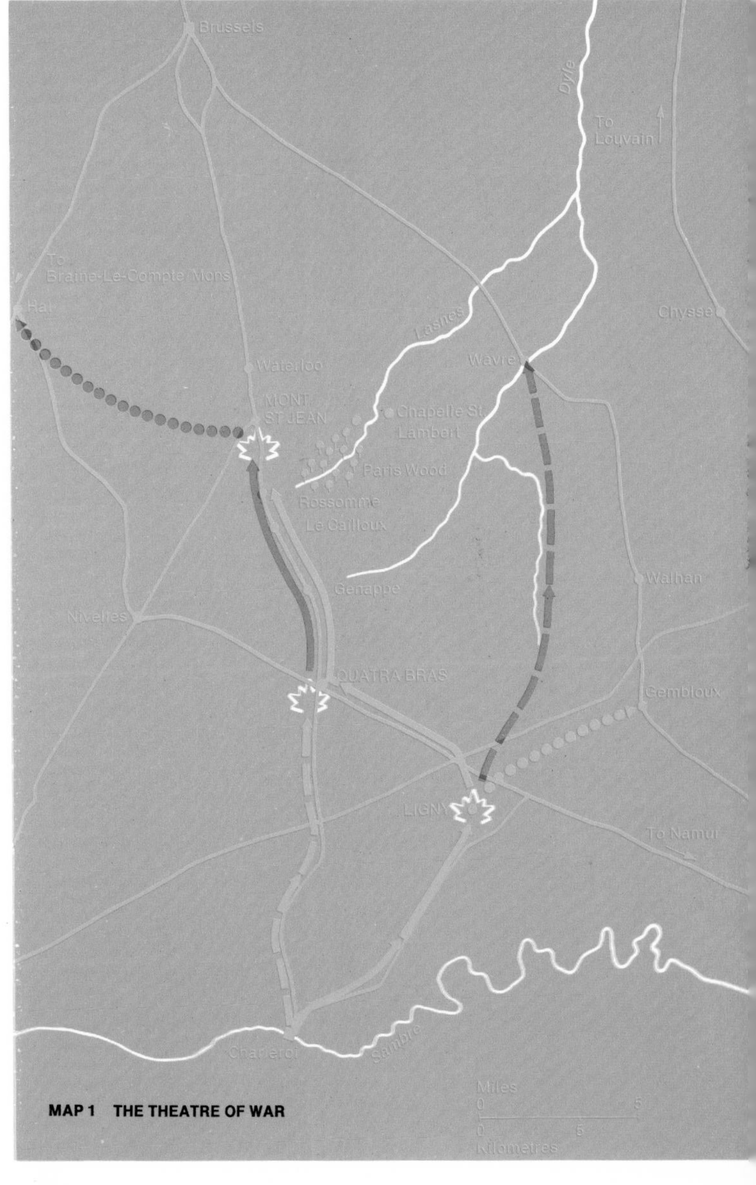

MAP 1 THE THEATRE OF WAR

Brussels. Blücher had succeeded in concentrating three of his four corps at Ligny, but by the evening his army had suffered a heavy defeat and was retiring in confusion. Blücher himself, unhorsed while leading a desperate cavalry charge, had been badly bruised. To the west Ney had encountered elements of Wellington's army at Quatre Bras; a bloody combat had developed and he failed to break through. By that evening Wellington had most of his army in hand near the crossroads. During the day he had received only one rather garbled report from Blücher, and when the firing died down around Ligny he assumed that the French had been repulsed.

But the next day he had an unpleasant surprise. Cavalry patrols revealed Ligny to be held by the French, and it was reported that Blücher, having been soundly beaten, had retired during the night towards Wavre. Exposed to a frontal attack by Ney, and an encirclement round his right and rear by the remainder of the French under Napoleon, Wellington's army, as he himself might have phrased it, was 'in a scrape'. He ordered an immediate withdrawal to Mont St Jean, about eight miles west of Wavre.

A somnolent Ney, licking his not inconsiderable wounds from the previous day, made little attempt to interfere, and save for a brisk cavalry action at Genappe, Wellington broke clean away.

While Napoleon, as was his custom, brooded over the battlefield of the previous day, he received the electrifying news from Ney at Quatre Bras that he had Wellington with most of his army in front of him. Despatching Grouchy with 30,000 men to follow up Blücher, the Emperor at once marched to join his Marshal. But it was too late; he had to content himself with following Wellington along the road to Brussels. In the early afternoon the skies opened, and for the rest of the day his soldiers trudged forward in heavy, drenching rain. That evening a strong British position was discovered astride the road just north of Mont St Jean. Napoleon bivouacked for the night, certain he could bring Wellington to battle on the morrow. To the east, Grouchy had halted for the night by Gembloux.

Wellington had taken up his new position confident that Blücher would join him next day. The nearest Prussians were little more than seven miles away. If they marched at dawn their leading regiments could have their breakfast at the Mont St Jean position. If the French attacked him, they would be crushed between the two armies like a walnut between the jaws of a nutcracker. It seemed far more probable that Napoleon would slip westwards round his right flank and interpose himself between the Allies and Brussels. Wellington directed Prince Frederick of the Netherlands with his corps and Colville's 4th Division, less Mitchell's brigade – in all about 15,000 men – to occupy and block the Mons road to Brussels.

Over by Wavre Blücher ordered the corps of Bülow and Pirch to march at daybreak for Mont St Jean. He proposed then to follow later with the rest of his army.

Marshal Gebhard von Blücher
Commander-in-chief,
Prussian Allied forces
Aged seventy-two at Waterloo, Blücher was by far the oldest of the three commanders. He was a magnificent leader with tremendous personality and powers of command. In battle, once fighting had begun, he also showed a notable tactical flair. He laid no claims to intellectual eminence – he left that to his staff – and in tactical and strategical ability he could match neither Napoleon nor Wellington; but for sheer courage and determination he was second to none. His hold over his troops was absolute. He died in 1819.

The Emperor Napoleon I

French Commander-in-chief

Since Austerlitz Napoleon's battles had become bloodier and more stubbornly contested. This was not through any lessening of his genius, but rather because under his tutelage his opponents were becoming more skilful and their armies more effective. Even so, in 1813 the Allied military commanders tried to avoid the armies under his personal command and concentrated on those commanded by his marshals.

At the time of Waterloo Napoleon was nearly forty-six years old. In 1814, although ultimately he failed, he fought one of the most brilliant campaigns of the age against the Allies, gaining remarkable victories at Champaubert, Montmirail, Vauchamp and Montereau. During the Waterloo campaign, despite claims to the contrary, it is evident that his genius was unimpaired. But never before had he to contend with adversaries so formidable; he had never met Wellington in battle, and the old war-horse, Blücher, had assimilated some of the lessons that Napoleon had taught him in earlier encounters. After Waterloo Napoleon was sent to his final place of exile on St Helena, where he died in 1821.

The Duke of Wellington

Commander-in-chief,
Anglo-Dutch Allied forces

The Duke of Wellington (1769–1852) was one of the greatest of British military commanders. He first achieved fame in India at the battles of Argaum and Assaye. Between 1808–14, with armies often vastly inferior in strength, he had first forced the French out of Portugal and then, in 1813, driven them out of Spain.

The French generals in the Peninsula, while admitting that he was unsurpassed at fighting a defensive battle, considered him cautious and a little rigid in his strategical concepts. However, without question he was a master of the battle-field and boasted with reason that he could command a company or an army with equal facility. He lacked the glamour, the aura of genius, of his great French rival; but thanks to his cool dispassionate logic and almost infallible instinct for the realities of a situation, he was never to know defeat. He had his forty-sixth birthday in May 1815, and during the Waterloo campaign was unquestionably at the height of his powers.

left Members of Napoleon's Imperial Guard, an elite force set up by the Emperor. Entry into the Guard was strictly controlled and pay and conditions – symbolized by uniforms of the greatest splendour – were superior to those found elsewhere in the French Army.

right Wellington arrayed his army along a 3,200 yard front between the Château of Hougomont and the Papelotte farmhouse, using a series of small ridges to conceal many of his dispositions from the French. Paramount in his mind was the need to guard against an attack directed at or round his right flank. In fact, however, Napoleon planned a crushing blow along and to the east of the Charleroi road. This was to be preceded by two diversionary attacks – one against Hougomont, the other to the east of Papelotte.

ANGLO-DUTCH

Cavalry
Infantry

1 Chassé
2 Garrison of 1st Guards, Nassau battalion and Hanoverians
3 Byng
4 Maitland
5 Colin Halkett
6 Kielmansegge
7 Ompteda
8 Mitchell
9 Duplat
10 Adam
11 Hew Halkett
12 Grant
13 Dornberg
14 Arentschild
15 Somerset
16 Kruse
17 Brunswick
18 Hanoverian
19 Belgo-Dutch
20 Lambert
21 Bijlandt
22 Nassau (Prince Bernhard)
23 Kempt
24 Pack
25 Vincke
26 Best
27 Vandeleur
28 Vivian
29 Ponsonby

FRENCH

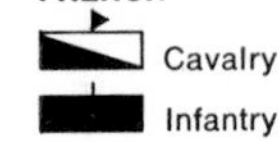

Cavalry
Infantry

1–4 Reille's Corps
1 Piré
2 Jérôme
3 Foy
4 Bachelu
5 Kellerman
6 Guyot
7 Lobau
8 Imperial Guard
9–12 D'Erlon's Corps
9 Allix
10 Donzelot
11 Marcognet
12 Durutte
13 Jacquinot
14 Milhaud
15 Lefèbvre-Desnouettes

The Rival Armies

The strengths of the respective armies on the morning of 18 June were approximately as follows:

Allies

a) Wellington's Anglo-Dutch Army at Mont St Jean

Infantry

British and King's German Legion	20,000
Hanoverian	10,000
Brunswicker	5,000
Belgo-Dutch	16,000
	51,000

Cavalry

British and King's German Legion	8,000
Hanoverian	1,000
Brunswicker	600
Belgo-Dutch	3,000
	12,600
Artillery train and Staff	4,500

Total 68,100 men (say 68,000) and 156 guns

b) Anglo-Dutch Corps at Hal	15,000
c) Blücher's Prussian Army at Wavre	
1st Corps (Ziethen)	20,000
2nd Corps (Pirch)	20,000
3rd Corps (Thielmann)	17,000
4th Corps (Bülow)	28,000

Total 85,000 men and 200 guns

French

a) Napoleon's Army bivouacked between Rossomme and Genappe

Infantry	53,000
Cavalry	16,000
Artillery train and Staff	5,000

Total 74,000 men and 240 guns

b) Grouchy's Corps at Gembloux 30,000

Grand total, Allies 168,000 men and 356 guns
Grand total, French 104,000 men and 240 guns

Note. Present strengths at the best of times are inaccurate and often differ markedly from ration strengths. It is probable that the armies were weaker than the figures shown, which are, particularly as far as the Prussians are concerned, mere estimates. Numbers have been approximated to the nearest 1,000.

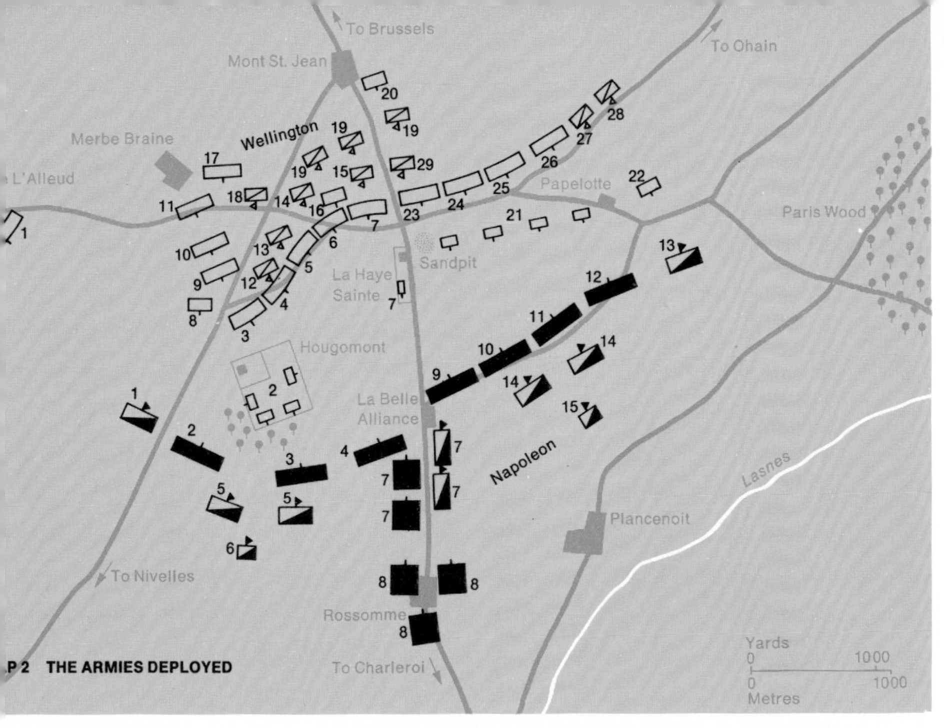

below The armies at Mont St Jean on the morning of 18 June. They are seen from the French side looking across the roof of La Belle Alliance towards the Château of Hougomont and the right-hand or western side of the Anglo-Dutch position.

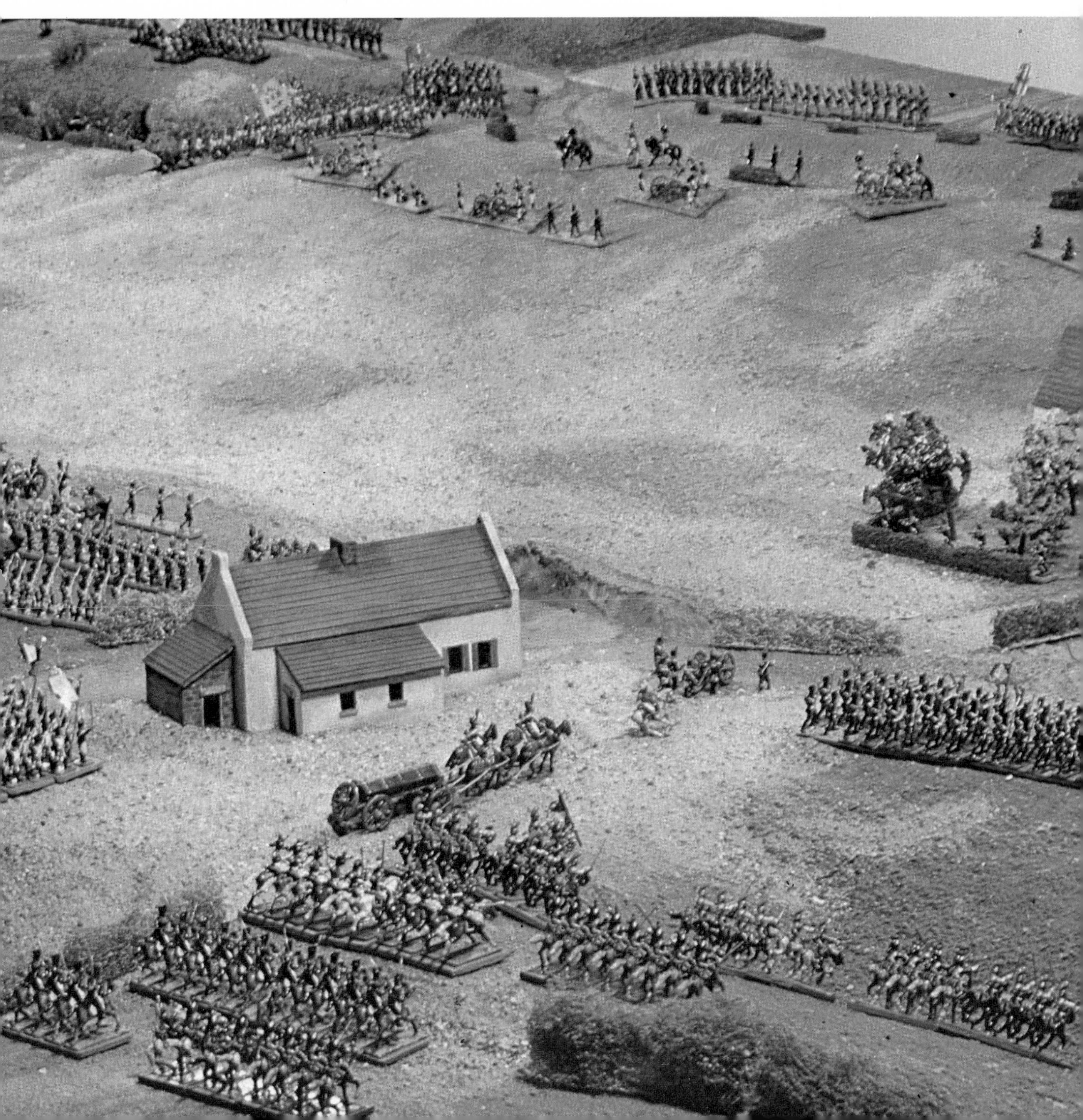

Wellington had spent the night at the little village of Waterloo. He does not seem to have slept well, for at three in the morning he was writing letters and at six had ridden forward to inspect the positions held by his damp and bedraggled troops. These lay along a series of ridges, little more than rolls in the ground, which ran roughly east and west, just north of Mont St Jean. Except for the rain-sodden soil and the standing crops which covered most of them, the ridges presented little obstacle to movement. Three important roads traversed the position. The Ohain road, slightly sunken and with low hedges to the east, ran generally below the crest and parallel to the front, the Charleroi road running north and south divided it into two nearly equal halves, and the road from Nivelles came in on the British right to join the Charleroi road by Mont St Jean. Jutting out from the line of ridges along which his army was aligned were three strong points that Wellington had occupied to serve as bastions to his main position.

It was apparent to Wellington that if the French attacked at all, and this appeared unlikely, their attack would almost certainly be directed round his right or western flank. Should they be so ill-advised as to attack his left, the Prussians driving in from Wavre would speedily dispose of them. Napoleon's most likely manoeuvre, he concluded, would be to envelop his right flank, probably with one corps, and thrust on to Brussels with the remainder of his army. Wellington had already planned against such an eventuality by stationing a strong force at Hal. Even so he was careful to dispose the rest of his heterogeneous and unwieldy army so that such a move could be met with the minimum of manoeuvre.

The French army had bivouacked between Rossomme and Genappe, many units not reaching their bivouac areas until after dark. At eight o'clock that morning the regiments started defiling to the right and left of the Charleroi road. Napoleon was convinced that the Anglo-Dutch army must be smashed where it stood, and to this end he planned a tremendous blow along and to the east of the Charleroi road.

At 11 a.m. his orders were prepared. To the west of the Charleroi road, Reille with his corps consisting of one cavalry division (Piré) and three infantry divisions (Prince Jérôme, Foy and Bachelu) was to mount a diversionary attack on Hougomont and cover the left flank of the main assault. This was to be made by d'Erlon with his corps of four infantry divisions, those of Allix,[1] Donzelot, Marcognet and Durutte, with the aim of capturing the crossroads at Mont St Jean. As a diversion, his light-cavalry division (Jacquinot) was to demonstrate east of Papelotte at the same time as Reille's men attacked Hougomont.

While the French were forming up for the attack, the Prussians were on the move. Bülow's corps, 28,000 strong, had been selected to lead the advance. It was the furthest away from Mont St Jean and well to the east of Wavre. Bülow had set off at seven that morning and by eight o'clock the head of his column was negotiating the narrow street that ran through Wavre. A house caught fire and the road was then blocked. It was not until 10 a.m. that Bülow's troops began to emerge. The Prussians were going to be late. At Gembloux, twelve miles to the south, Grouchy broke camp and resumed his march on Wavre. The actors were beginning to take their places for one of the greatest of military dramas.

The Course of the Battle

Punctually at 11.30 a.m. Prince Jérôme's division began to advance on Hougomont. The sun had yet to penetrate an overcast sky when, with bugles sounding and the drums beating forward, Jérôme's leading brigade plunged into the copse and orchard that surrounded the Château. The Nassau battalion took the first impact; filtering back through the trees they took a heavy toll of their assailants, but could not withstand their spirited advance. The French forced their way to the edge of the copse, but here an unpleasant surprise awaited them. The garden of the Château was surrounded by a solid six-foot masonry wall; through it and over it protruded the muskets of the four light companies of the Guards; they had loopholed the walls and constructed firing platforms so that they could fire over the top. Only thirty yards separated the copse from the wall, yet in the face of the deadly fire of the defenders, it became an impossible gap to cross.

Seated on his horse, Wellington watched the combat from the ridge behind. As he had expected, the French were attacking his right flank; but he was jumping to no conclusions while great masses of French infantry lay deployed on his front. With a niggardly hand he fed forward reinforcements to the Château from Byng's brigade, little more than a company at a time. Jérôme, furious at being checked, sent in more troops. French soldiers surged up to the Château, where a bitter struggle was to continue throughout the rest of the day.

Meanwhile preparations went ahead for the main assault. Over by Papelotte Farm Hacquinot with his cavalry had made a rather perfunctory demonstration, finding the broken ground unsuitable for horsemen. A great battery of eighty guns lined up east of La Belle Alliance and opened fire at about midday. At 1 p.m. Ney, who commanded the two corps of d'Erlon and Reille and was acting as field commander to Napoleon, rode up to the Emperor, seated on his horse on the top of a mound just north of Rossomme, to ask permission to begin the attack. The Emperor was looking eastwards, staring intently at a black stain that seemed to be spreading in front of the wood by Chapelle St Lambert, four miles to the east. All doubts were speedily resolved. A cavalry patrol reported that the Prussians had arrived at Chapelle St Lambert.

The Emperor ordered Lobau, with his two divisions and

[1] Allix was not present at the battle; his division has sometimes been referred to by the name of his senior brigade commander, Quiot.

the two divisions of light cavalry deployed beside him, to the area of Paris Wood to guard the right flank. Already, without firing a shot, Blücher had subtracted four divisions from the forces facing Wellington. Those around the Emperor felt a sudden chill, but he appeared unmoved. Soult, Napoleon's Chief of Staff, wrote a despatch to Grouchy telling him to 'manoeuvre towards the French right'. Grouchy, however, did not receive it until five o'clock that evening, so Soult might have saved himself the trouble. At midday, Grouchy at Walhain, some thirteen miles away, heard in the west the distant booming of a heavy cannonade. Gérard, one of his corps commanders, urged him to march 'to the sound of the guns'. But a local inhabitant had told Grouchy that the Prussians were mustering at Chysse well to the east of Wavre. Confident that he could interpose between the Prussians and Napoleon, Grouchy continued his march towards Wavre, where he eventually engaged Thielmann's corps.

Wellington, too, was viewing the moves of the Prussians, or the slowness of their moves, with some anxiety. It was apparent that he would have to meet the first shock alone, and the main attack was coming in on his weak left wing. Napoleon had deceived him again. To strengthen his flank he moved Ponsonby's Union brigade to the left behind Pack and brought Somerset's Household Cavalry brigade across the road to replace them.

At 1.30 p.m. the bugles sounded again and the drums began to beat. The blue-coated columns of d'Erlon's infantry came steadily forward. On the left, Allix's division advanced in two brigade columns with Travers' steel-plated cuirassiers walking their horses behind. To Allix's right at intervals of about three hundred yards, two enormous phalanxes of bayonets nearly two hundred yards across and eighty deep, each composed of some 4,000 tightly packed men, strode forwards. These were Donzelot's and Marcognet's divisions, every battalion in a three-deep line and ranged one behind the other with an interval of four yards between battalions; each column had a front of about 170 men and a depth of twenty-four ranks. Away to the right by Papelotte, Durutte had sent his division forward in battalion columns; each battalion marched with the companies strung one behind the other, the leading companies of each keeping roughly in line (Durutte had noticed the houses and broken country in front and had adopted a flexible formation).

As the columns approached, the British guns began to rake the packed ranks. But the men closed up and with a great shout of 'Vive l'Empereur' pressed on to the crest. On both flanks, by La Haye Sainte and Papelotte, a fierce combat developed. In the centre, Bijlandt's light infantry was rudely brushed aside and Donzelot's and Marcognet's divisions drove on over the top of the ridge. As they appeared, Picton led forward the 5th Division in line to oppose them. Soon from La Haye Sainte to Papelotte the roar of musketry became incessant and black clouds of smoke blanketed the ridge.

By La Haye Sainte the British companies of the 95th were driven out of the sandpit and the German riflemen of the King's German Legion compelled to seek shelter in the farmhouse; but here they stood, immovable as granite. The French infantry, swarming round the farmhouse, isolated it from the crest behind. Kielmansegge sent forward a battalion from his Allied Hanoverian brigade to help the garrison. Travers saw his chance, and swooped down on the unfortunate infantry with his cuirassiers and cut them to pieces; his horsemen then pushed on for the ridge. In the centre Picton was dead and his division hard pressed; on the left Durutte, despite a gallant resistance

As early as 1 p.m. the advance of the Prussians from Wavre compelled Napoleon to divert four divisions under Lobau to protect his right flank near Paris Wood. Then, as first Bülow followed by Pirch and Ziethen drove into the French lines, more and more French troops had to be dispatched to oppose them. By 6.30 p.m., however, the Prussians had arrived in such strength that Napoleon faced encirclement.

ALLIES

Anglo-Dutch

Prussians

1 Bülow 4.15 p.m.
1a Bülow 5.30 p.m.
1b Bülow 7.30 p.m.
2 Ziethen 6 p.m.
2a Ziethen 6.30 p.m.
2b Ziethen 7.30 p.m.
3 Pirch 6.30 p.m.
3a Pirch 7.30 p.m.

FRENCH

Cavalry

Infantry

Approximate positions

1 Lobau 2 p.m.
1a Lobau 7.30 p.m.
2 Young Guard 5.30 p.m.
3 2 Bns. Old Guard 6.30 p.m.
4 Durutte 6.30 p.m.
4a Durutte 7.30 p.m.

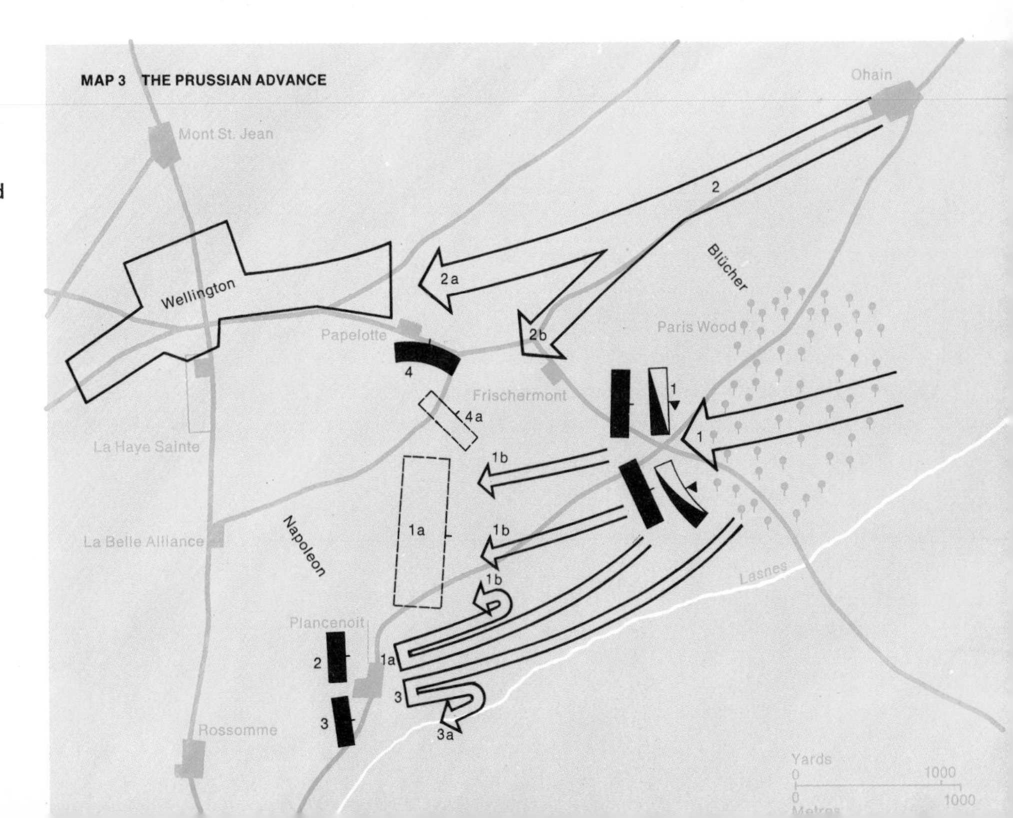

from the Nassau brigade, had captured Papelotte and his left-hand brigade was bearing down on Best's Hanoverians. The situation looked critical.

Wellington had meanwhile been watching the enemy closely. After the long march forward in a tight formation and the obstinate resistance that they had encountered, the leading ranks of the French had become disordered. With the same immaculate timing he had shown at the Battle of Salamanca, Wellington ordered the two heavy-cavalry brigades to charge.

By La Haye Sainte, the Household Cavalry caught Travers' troopers trying to negotiate the Ohain road. They smashed into the surprised cuirassiers and routed them; then they thundered down on Allix's division, broke into its ranks and sabred them mercilessly. To their left, Ponsonby's Union brigade had trotted forward. Pack's and Kempt's men opened their ranks to let them through, then the cavalry descended like a whirlwind on the French. As the Scots Greys passed through the ranks of the 92nd, the Gordon Highlanders raised a great shout of 'Scotland forever', and some charged forward a short distance with the horsemen.

It was one of the greatest of British cavalry charges. As the horsemen hurtled down upon them the two great infantry phalanxes of the French reeled, then fell apart. A berserk fury seized the cavalrymen of both brigades; sabring everyone in their path and heedless of bugle calls to rally they pounded on into the heart of the French position, and here catastrophe overtook them. West of the Charleroi road, Bachelu's right-hand brigade swung round and blasted them in the flank. In front Milhaud's cuirassiers charged forward, while on their left flank Jacquinot led in his lancers. Their ranks in chaos, their horses blown, the British horsemen could offer little resistance. The Household Cavalry, which had kept a reserve, escaped utter disaster, but the Union brigade was nearly exterminated.

By now it was nearly three o'clock on a hot, clammy afternoon and already the battle-smoke was beginning to hang low. Around Hougomont the relentless struggle continued. The whole of Jérôme's division had become embroiled and battalions from Foy's were being sucked into the battle. Byng's complete brigade, except for two companies left on the ridge with the colours, had joined the garrison of the Château, and at this juncture Napoleon directed a battery of howitzers to shell it. The Château and its outbuildings caught fire; the chapel, in which the wounded of both sides had been laid, blazed up and many were burned. But nothing could shake the iron resolution of the Guards.

Along the rest of the front the French guns opened a murderous bombardment. Napoleon still held to his intention to break the Allied centre along the line of the Charleroi road. He recognized that La Haye Sainte was the key and ordered Ney to capture it immediately. By about half-past three Ney had scraped together enough men to launch an attack. He led forward a brigade from Allix's division while Bachelu's men and a brigade from Donzelot shielded its flanks. The combat was short, bloody and fruitless. Before the steady fire of the German riflemen and the companies of the 95th, the assailants melted away.

Again there was a pause while the massed French guns thundered on. Except for that of the Guard, the French infantry had suffered severely, and less hardened troops might have abandoned the battle; it would take time for the disorganized regiments to reform, and their first keen ardour had evaporated. Away to the east, scattered firing could be heard as Bülow's Prussians, emerging from Paris Wood, found Lobau's men waiting for them on the far side. Now Ney took a desperate decision. Time was running out. He dared not wait for the French infantry to reform and his impetuous spirit was not given to waiting. He resolved to attack the Allied centre with cavalry alone.

At about 4 p.m., 5,000 horsemen punctiliously took up their dressing. The bugles sounded; the long ranks broke into a walk, then a trot. Despite the heavy and muddy

left French troops storm the gate of La Haye Sainte.

above A member of the Royal Scots Greys in Ponsonby's Union brigade engages a French infantryman. **left** The massed advance of Ney's cavalry, with La Haye Sainte in the background. Facing the horsemen on the reverse slope are the Anglo-Dutch infantry squares.

ground, the French cavalry rode steadily forward, rank on rank. The British guns opened rapid fire, roundshot then grape. As the horsemen breasted the slope the carnage was terrible. For a moment they faltered. Then, with a great cry of 'Vive l'Empereur', they galloped over the crest, over the guns and on to the silent scarlet squares beyond.

Wellington had seen the French cavalry form up with an incredulous wonder. He found it difficult to believe they would dare to attack unsupported by infantry. Without haste his army formed a succession of squares, one or two battalions in each. Alternate squares were echelonned back to avoid firing into each other; and the fire from the face of one could then cover the face of the other.

As the French cavalry approached, the gunners fired a last round and bolted into the squares for protection. Then the cavalry surged round them, vainly looking for an opening. But the two leading ranks of infantry formed an impenetrable hedge of bayonets, while from the two ranks behind rolled out a continuous and deadly musketry.

At this time the Duke was everywhere, taking refuge in a square, then, as the tide of horsemen receded for a few moments, darting to another, encouraging his men to hold firm and making hasty readjustments where danger threatened. Some of the Allied cavalry charged into the fray to relieve the pressure on the infantry, but these could make little impression on their armoured opponents, and some regiments refused to charge at all. As charge succeeded charge, the French cavalry moved round the squares, apparently masters of the ground, but unable to beat down the resistance of the stubborn British infantry.

While this fearful combat raged, the Prussians were advancing inexorably. Lobau's outnumbered French force might hold its foes in front, but Blücher, shrewdly taking advantage of his superior numbers, sent more men to turn Lobau's flank and rear. Bülow took the village of Plancenoit and Prussian guns began to shell the Charleroi road. Napoleon despatched Duhesme's division of the Guard to sustain Lobau and with great dash they retook the village. But the afternoon was passing; it was now nearly 5 p.m. and over on the ridge Milhaud's shattered squadrons had begun to abandon their hopeless task of breaking down the Allied squares.

It was vital that Wellington should be destroyed before the Prussians could deploy their full strength. Napoleon ordered Kellerman with his corps into the inferno on the ridge. Buyot and his Guard cavalry followed. Once more the awful struggle of the horsemen and the squares was resumed. Ney led the charge. He raged round the battlefield like a god of war. Horse after horse had been shot from under him, and at one time he was seen in his fury belabouring an abandoned British gun with his sabre.

Heavy clouds of gunsmoke shrouded the ridge, but the cavalrymen looming through the fog could not break into the slowly thinning squares. Ney went back to bring forward such infantry as were still capable of fighting. In time French skirmishers began to get a footing all along the ridge. But Ney had not sufficient men to exploit his advantage, and those he had were exhausted by the Herculean efforts they had already made. He sent back urgently to Napoleon for more men.

But the Emperor had none. The Prussians had renewed their attacks. Duhesme's division of the Guard had been driven out of Plancenoit in disorder. A bare two thousand yards now separated the Prussians from Hougomont. The French faced encirclement. Napoleon disposed his two remaining divisions of the Guard in squares from Rossomme to La Belle Alliance. Two battalions of the Old Guard he despatched against Plancenoit. Miraculously, the two superb battalions of veterans retook the village and Duhesme's division rallied. The situation had been restored, but the last chance of a French success had vanished.

By Mont St Jean the French infantry attacks petered out. The infantry squares still stood immovable and Kellerman's broken squadrons were beginning to drift away from that fatal ridge. Now Wellington brought forward every soldier he could find to the centre of his position. Under a heavy fire he personally led forward the Brunswickers; by Papelotte, Ziethen's corps had begun to arrive and Durutte was now fiercely engaged with the Prussians. Wellington drew in Vivian's and Vandeleur's brigades to his centre and Vincke from the left of Pack. Chassé's Belgo-Dutch division he had already brought from Braine l'Alleud to Merbe Braine and now he pushed it forward. Adam's brigade from the 2nd Division came up on the left of Maitland's Guards. A steady line was reformed and the French skirmishers were thrown off the ridge. Only by La Haye Sainte a gallant few from Donzelot's and Allix's divisions maintained a precarious foothold.

The day was drawing on, the sun, a blood-red orb, was declining over Braine l'Alleud, its rays no longer obscured by cloud, occasionally piercing the rolling smoke that swirled across the battlefield. And all the time the guns rumbled on. Napoleon must have known the battle was lost, but he resolved on one last gambler's throw, a final act of daring: 'De l'audace et toujours de l'audace'. A miracle might happen.

The division of the Middle Guard was intact and so was the Old Guard except for the two battalions at Plancenoit. For the moment the Prussians were held. Napoleon took the Middle Guard forward and handed them over to Ney for a final assault. Of the Old Guard, he left the 1st battalion of the 1st Chasseurs, his personal bodyguard, at Le Cailloux. The remaining five battalions he kept in hand to exploit a breakthrough or stave off a disaster.

It was by now nearly 7.30 p.m. and Pirch's and Ziethen's corps were coming in on the French right in strength. To stop an incipient panic, Napoleon spread the word that Grouchy had arrived.

Wellington quietly prepared for the last act of the drama. A deserter had warned him that the Guard would attack in half an hour. Duplat's brigade of the King's German Legion had joined Byng's Guards brigade in Hougomont. Wellington had Hew Halkett's brigade holding the ridges behind, then came Maitland's Guards and Colin Halkett's battered brigade; he had ringed La Haye Sainte with Brunswick and Hanoverian troops. The 6th Division was astride the Charleroi road. Pack and Kempt continued the line to the east where the Prussians of Ziethen were arriving in ever-increasing numbers. He had placed Vivian's and Vandeleur's cavalry brigades in the centre as a reserve, and behind Colin Halkett's brigade he placed Chassé's Belgo-Dutch division which had hardly been engaged. Behind Adam and Maitland he sited Mitchell's brigade, also comparatively untouched.

His dispositions completed, Wellington waited for the final shock. He then instructed his regiments to form their line four deep and to take what cover they could from the fire of the French guns, and rode over to where Maitland's brigade of Guards were lying down in the corn; there, he could see, the heaviest attack was going to fall.

As the light was slowly fading over the smoke-filled battlefield, Ney led the Guard forward. The scene had the fated grandeur of the last act of a Greek tragedy. The Guard marched to their last attack in impeccable order, their drums beating them forward, their generals at their head. As the column advanced the heads of each battalion swung away to the left in succession, so that they became echelonned nearly from La Haye Sainte to Hougomont like five steps of a staircase; between the battalion columns horse artillery guns took station; on the right and in front of all the others were the 1st battalion of the 3rd Grenadiers, the choicest of the choice.

What happened as the columns of the Guard broke through the battle smoke and burst in upon the British line can never now be known for certain. It seems that the 1st/3rd Grenadiers, some troops from Donzelot's division and perhaps part of the right-hand company of the 4th Grenadiers drove in some Brunswickers and forced Colin Halkett's men back. Halkett clutched a colour and rallied his men; he was severely wounded, but Chassé brought forward a battery on Halkett's right and deployed Dittmer's brigade, 3,000 strong, on his left. Swamped by numbers, the French were nearly annihilated.

Before Maitland's brigade of Guards the 4th Grenadiers and the 1st/3rd Chasseurs suddenly appeared over the crest. The French raised a shout of triumph, believing they had broken through. 'Now, Maitland,' said the Duke urgently. 'Now is your time.' The 2nd/1st Footguards and the 3rd/1st Footguards (now the Grenadiers) rose to their feet. A scarlet rampart confronted the astonished Frenchmen, a rampart which belched flame and destruction. For a minute or so a savage musketry battle raged. The French guns may have managed to fire a few rounds of grape, but the Duke saw that the ranks of the French were becoming unsteady, and he ordered the Guards to charge. The long rows of bayonets flickered down and the red line swept forward; the Grenadiers and Chasseurs, what was left of them, ran back down the slope. Then, after a pause they rallied when they encountered the 2nd/3rd and 4th Chasseurs coming up the slope, and all four battalions flung themselves against Maitland's and Adam's brigades.

It was a gallant but hopeless assault. From in front they were met by a blinding assault; for a few moments the battalions of the Guard stood firm and exchanged a deadly musketry fire, then they suddenly collapsed, and the slopes were dotted with groups of men running back.

All round the battlefield the anguished cry rang out, 'La Garde recule'. Their regiments shattered, and with 60,000 Prussians coming in on their flank and rear, the French army began to disintegrate. Vivian's and Vandeleur's brigades of light cavalry swooped down on the retreating Middle Guard and transformed them into a mass of helpless fugitives. Wellington rode along his line telling his weary and incredulous soldiers to advance. All over the battlefield French soldiers were throwing away muskets and equipment in a mad scramble to escape. Only the Old Guard preserved their order and retired slowly in square. Allied cavalry surrounded them, then came the infantry and guns. In the gathering darkness the squares of the Old Guard were shot through and through. British officers, appalled by the senseless slaughter, called on them to surrender. Cambronne, commanding the 1st Chasseurs, refused and the desperate combat continued until the squares were demolished.

Napoleon, who had taken refuge in a square of the Old Guard, left the field almost fighting his way through lawless hordes of soldiers, while at nine o'clock that night Wellington and Blücher met near the aptly named inn of La Belle Alliance.

Aftermath and Conclusion

The cost of the battle was frightful. The Allied army lost about 15,000 men of whom about 7,000 were British; the Prussian casualties were about the same as the British. The casualties of the French are difficult to assess; their losses must have been between 20,000 and 30,000 men, but the army as such ceased to exist.

What would have happened if Napoleon had won the Battle of Waterloo can only be a matter for conjecture; he still had powerful enemies to contend with. It is safer to recount what followed. At Wavre Grouchy broke off his battle with Thielmann and conducted a skilful withdrawal. But Napoleon's situation was hopeless. He abdicated, fled to Rochefort and surrendered, appropriately enough, to his most implacable foe, the Royal Navy. He sailed in HMS *Bellerophon* to England and thence to St Helena, exile and death in 1821. The Allies occupied Paris, and then negotiated a settlement which brought Europe peace for thirty-three years, and freedom from a continental conflagration for all but a century.

Clifford C. Johnson at

GETTYSBURG

This was the greatest battle of the American Civil War. After three days of savage fighting Meade's Union army beat off Pickett's charge, the last of Lee's furiously sustained offensives. The Confederate army, at last crippled by its losses, was compelled to retreat. From that moment the war swung irreversibly to the Union side.

1863

A head-on view of Buford's Union cavalry division barring the Confederate advance on Gettysburg.

The Background to the Battle

In the late spring of 1863 the leaders of the Confederacy were presented with a strategic problem of great magnitude and immediacy. Their problem, put simply, was to break the noose that was gradually tightening around the South while their armies were still capable of it.

The Confederacy was now everywhere on the defensive, and despite striking tactical successes in the east, the overall situation was bleak. There was only one army in all the Confederacy still capable of producing a decisive strategic victory. That was Lee's Army of Northern Virginia.

By late May, Lee had decided that the existing defensive policy would yield nothing and he resolved upon an invasion of the North. Lee's plan called for the Army of Northern Virginia to swing to the west behind a cavalry screen, thrust quickly up the Shenandoah Valley behind the curtain of the Blue Ridge Mountains, and debouch into the fertile Cumberland Valley of Pennsylvania. This daring operation would transfer the war from prostrate, wasted Virginia and compel the Union Army of the Potomac to march northward to defend the cities of Harrisburg, Philadelphia, Baltimore and Washington, DC.

There was negligible opposition to the plan within the Confederate cabinet, and on 3 June 1863 the first of Lee's divisions slipped stealthily from the line of the Rappahannock at the head of the army's march to the Shenandoah Valley. The great invasion of the North had begun.

Stuart's Raid

Lee's fine Cavalry Corps was commanded by the legendary Major-General J. E. B. ('Jeb') Stuart. Both physically and intellectually, Stuart was surely one of the most gifted horse-soldiers of all time. Stuart's men, who used to refer to themselves as the Southern 'chivalry', had time and again beaten the Yankee cavalry in a series of audacious and brilliantly conceived raids.

In the coming campaign, Lee would need Stuart's men to screen his march, to scout and follow the enemy's plans and dispositions, and to *keep him informed*. The cavalry was to act as the 'eyes and ears of the army', as Lee put it.

This was certainly not the time for Stuart to go off on one of his patented raids on the rear of the Army of the Potomac. Yet from 25 June to 2 July Stuart and his men were doing just that. In consequence, virtually all the information Lee had about the movements of the Army of the Potomac emanated from one Harrison, a scout attached to Lieutenant-General Longstreet's 1st Corps.

What had prompted Stuart to go off on this raid at such a crucial time? Clearly the fault lay in some measure with Lee, whose orders to Stuart had been general enough to allow Stuart to believe that such a raid was permissible. But more weight must be given to the theory that Stuart's notorious vanity had been wounded by a surprise attack at Brandy Station, Virginia on 9 June. There Stuart's men had been roughly handled by a large task force of Union cavalry

The three-day Battle of Gettysburg took place on 1–3 July 1863. Following his victory at Chancellorsville in early May General Lee determined to invade the North. On 3 June his Army of Northern Virginia began its march from Fredericksburg, swinging away from the Union lines and thrusting through the Shenandoah Valley into Pennsylvania. Mistakenly, however, Lee allowed his cavalry under 'Jeb' Stuart to head east to raid the rear of the Union army, thereby depriving himself of valuable support at a critical time (25 June–2 July). On 28 June Lee learned that the Union army had crossed the Potomac in pursuit, and he ordered his scattered corps to concentrate at Cashtown, nine miles west of Gettysburg.

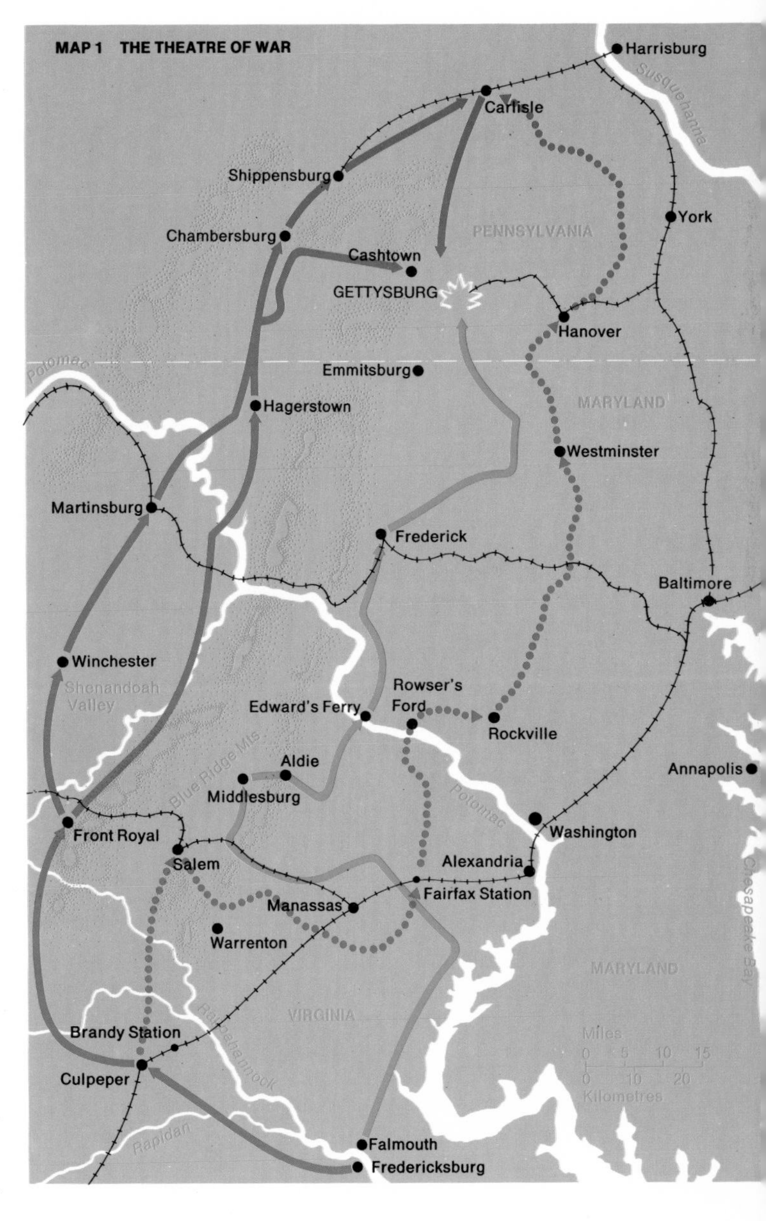

On these pages is a collection of military scenes recorded in the first age of photography. **far left** A Louisiana Zouave of the Confederate Army. Zouave regiments based on French models sprang up on both sides in the American Civil War, especially on the Confederate side among the former French colonies. **far left, below** A review of Union troops at Cumberland Landing towards the end of 1862. **left** A 6-pounder Welaro gun in the Union artillery. **right** Scouts and guides to the Army of the Potomac (the Union force that Meade was to command at Gettysburg). **centre right** A picket station returns the enemy's fire. **far right** A Union trooper awaits his orders, February 1863. **far right, below** A Confederate battery at Pensacola Bay, Florida, where old-style guns were used dating from 1812.

under Major-General Alfred Pleasanton. Quite possibly Stuart hoped the present raid would retrieve his reputation.

On the evening of 28 June Lee learned from Harrison that the Union army was north of the Potomac River. This was his first hint that the enemy had reacted vigorously to his march and was following closely. He ordered his scattered army corps to concentrate at Cashtown, Pennsylvania, nine miles west of the market-town of Gettysburg.

The Rival Armies

The comparative strengths of the rival armies on the morning of the first day at Gettysburg were approximately as follows:

Union		**Confederate**	
Infantry	73,000	Infantry	54,000
Cavalry	13,000	Cavalry	12,500
Artillerymen	7,500	Artillerymen	6,000
Total		*Total*	
93,500 men and 370 guns		72,500 men and 287 guns	

The Union army was organized under the command of Major-General George G. Meade in seven corps, each comprising three divisions (except the 3rd and 12th Corps which had two divisions), together with a cavalry corps (three divisions and two brigades of horse artillery) and an artillery reserve. Meade's corps commanders at Gettysburg were as follows:

Major-General George G. Meade
Union Commander-in-chief
Major-General Meade (1815–72) assumed command of the Army of the Potomac just three days before the Battle of Gettysburg. Introspective, apologetic and more precisely intellectual than Lee, Meade was not the man to excite the emotions of the rank-and-file. He looked like a village schoolmaster, was prone to nervousness and sometimes let his violent temper get the best of him – so much so that the men dubbed him 'Old Snapping Turtle'. But there was little grumbling when he assumed command; Meade had a reputation as a solid combat officer and had won preferment for the command over several other senior officers.
The new commander himself was philosophical: 'I may fail in the grand scratch; but . . . I shall probably do as well as most of my neighbours.' On learning of Meade's appointment, General Lee made a more objective estimate of his new adversary: 'General Meade will commit no blunder in my front, and if I make one he will make haste to take advantage of it.'

General Robert E. Lee
Confederate Commander-in-chief
In June 1863, General Robert E. Lee (1807–70) was the confident leader of an army he regarded as invincible. These were men who could 'go anywhere and do anything if properly led', and Lee had provided brilliant leadership ever since he had assumed command of the army just over a year before. The string of victories gained against formidable odds had been impressive and had undeniably earned Lee a prominent place among the great captains of history.

There is very little about Lee that does not command the admiration of men. He was a noble figure, revered by his men and feared and respected by his enemies. His failures were not those that might easily affect the outcome of a campaign or battle. He did allow his corps commanders a certain latitude or discretion in carrying out his orders, but they had shown themselves capable of independent judgment, and Lee was confident in them.

1st Corps Major-General John F. Reynolds (killed); subsequently Major-General Abner Doubleday
2nd Corps Major-General Winfield S. Hancock
3rd Corps Major-General Daniel E. Sickles
5th Corps Major-General George Sykes
6th Corps Major-General John Sedgwick
11th Corps Major-General Oliver O. Howard
12th Corps Major-General Henry W. Slocum
Cavalry Corps Major-General Alfred Pleasanton
Chief of Artillery Brigadier-General Henry J. Hunt

After his victory at Chancellorsville, General Robert E. Lee had reorganized his Army of Northern Virginia into three corps, under the following commanders:

1st Corps Lieutenant-General James Longstreet
2nd Corps (formerly that of 'Stonewall' Jackson) Lieutenant-General Richard S. Ewell
3rd Corps Lieutenant-General Ambrose P. Hill

Each corps was made up of three divisions; one battalion of artillery was assigned to each division and one or more battalions of reserve artillery to each corps. The Cavalry Corps remained under Major-General J. E. B. Stuart, and the artillery was commanded by Brigadier-General W. Pendleton.

The core of each army was its infantry. The men were nearly all armed with the rifled musket, which had been adopted by all western armies as the standard infantry firearm. There were a few breechloaders about, and, especially in the Union cavalry, an entire unit might be armed with them. These were very efficient weapons, but there were not many of them. Some Confederate cavalrymen maintained their individuality by using sawn-off shotguns in close combat! The bayonet was used by most infantrymen as an entrenching tool or cooking instrument, and wounds inflicted with this weapon were extremely rare.

Rifling – the cutting of grooves on the inside of a barrel the better to control a projectile's path – had improved the capabilities of the artillery. Many gunners, however, especially among the Confederates, preferred to use the bronze

below The advance of Heth's Confederate division on Gettysburg during the first morning of the battle.

Napoleon smooth-bore gun-howitzer rather than the newer rifled guns. When firing canister at 800 yards, the former was an awesome weapon of defence. The artillery also fired solid shot, shell, spherical case (shrapnel) and an incendiary device known as a 'carcass'.

Gettysburg: The First Day, 1 July 1863

On 30 June, Brigadier-General Johnston Pettigrew's North Carolina brigade of A. P. Hill's Confederate 3rd Corps approached Gettysburg in quest of shoes (a large stock was rumoured to be in the town). Pettigrew's men had a skirmish with Union cavalry to the west of Gettysburg. Some of his officers thought they heard infantry drums beyond the town, and Pettigrew withdrew prudently to Cashtown.

Opinion at Confederate headquarters was that the Union army was still too far away to prevent the Confederates from getting their shoes, but on the basis of Harrison's information Lee felt that it would be best to concentrate at Gettysburg. At 5 a.m. on 1 July Major-General Harry Heth's division, followed by Major-General Dorsey Pender's, set out for Gettysburg. Three miles west of Gettysburg, on the Chambersburg Pike, they encountered Brigadier-General John Buford's Union cavalry division barring the way to the town – as it had the day before. For two hours Buford's troopers, who were armed with newly issued Spencer seven-shot repeating carbines, held up the Confederate advance.

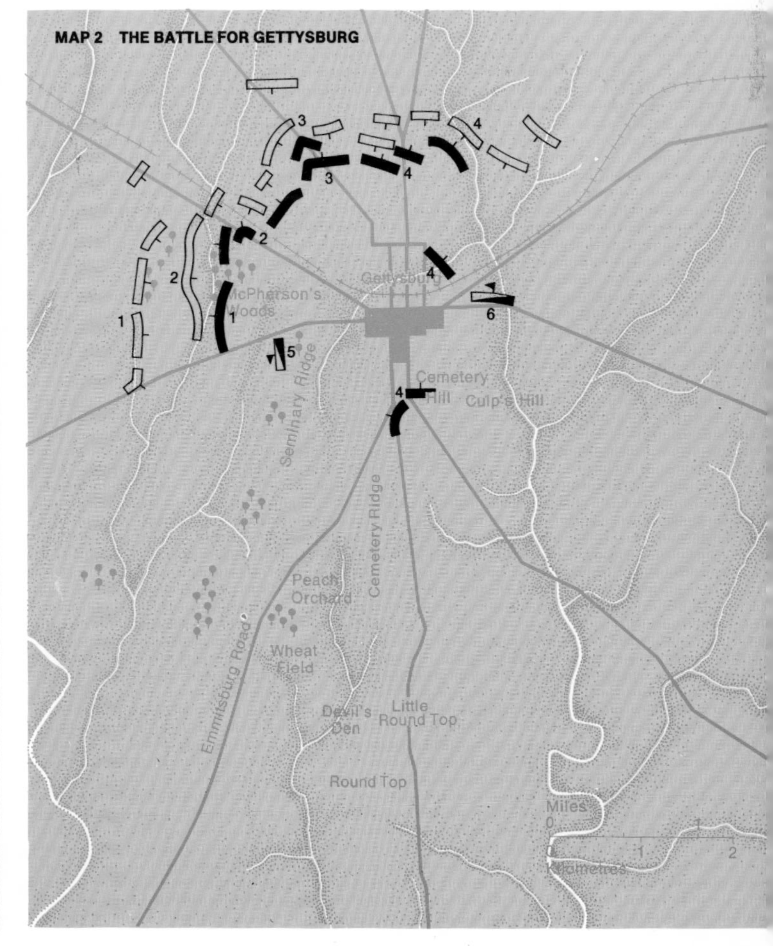

The map shows the positions of the leading forces of both sides at 2.30 p.m. on the first day (1 July). Early's Confederate brigade, driving in from the north, engaged and subsequently broke the Union 11th Corps, which retreated in disorder towards Gettysburg. On the Union left Doubleday's hard-pressed 1st Corps began a fighting withdrawal, aided by Gamble's brigade of Buford's cavalry.

UNION
Army Corps
Cavalry Corps
1-3 Doubleday (1st Corps)
1 Rowley
2 Wadsworth
3 Robinson
4 Schurz (temporarily commanding 11th Corps)
5-6 Buford (Cavalry Corps)
5 Gamble
6 Devin

CONFEDERATE
Army Corps
1-2 A P Hill (3rd Corps)
1 Pender
3 Heth
3-4 Ewell (2nd Corps)
3 Rodes
4 Early

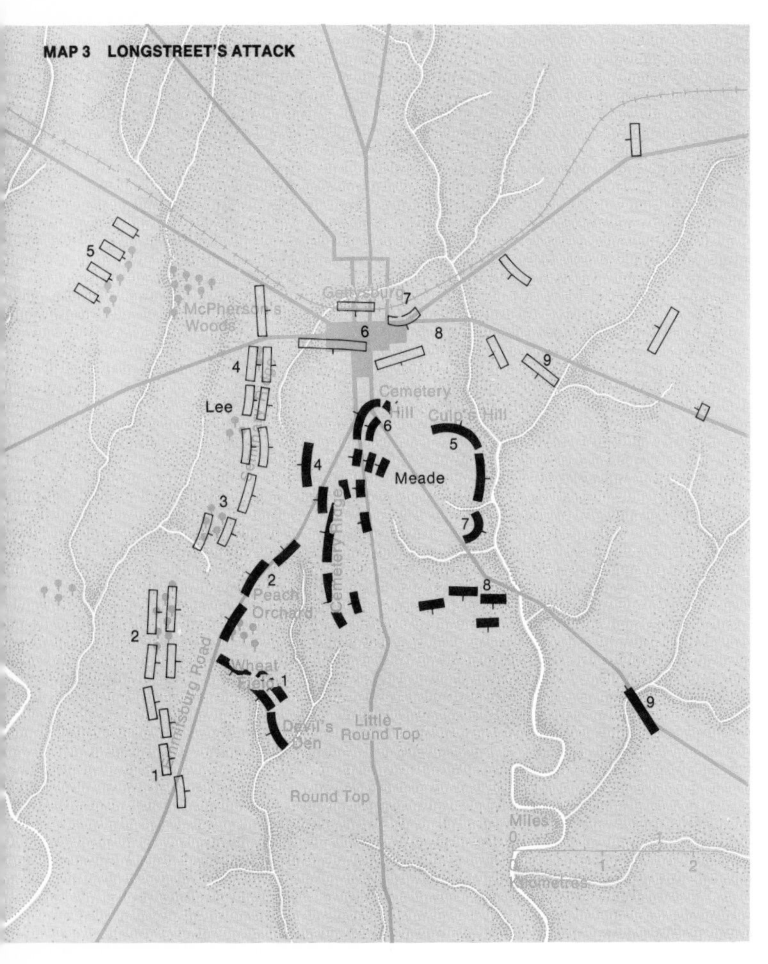

Early on the second day Meade prepared to defend a strong position on high ground between Little Round Top to the south and Culp's Hill to the north-east. At 3.30 p.m., after various delays, Longstreet's Confederates were poised to attack the Union left, which Sickles had weakened by moving his 3rd Corps forwards from its morning position on Cemetery Ridge. Sickles was now dangerously exposed to Longstreet's assault; he had also made the serious error of removing his troops from the vital Round Top hills.

UNION
Army Corps
1-2 Sickles (3rd Corps)
1 Birney
2 Humphreys
3 Hancock (2nd Corps)
4-5 Doubleday (1st Corps)
4 Newton
5 Wadsworth
6 Howard (11th Corps)
7 Slocum (12th Corps)
8 Sykes (5th Corps)
9 Sedgwick (6th Corps)

CONFEDERATE
Army Corps
1-2 Longstreet (1st Corps)
1 Hood
2 McLaws
3-5 A P Hill (3rd Corps)
3 Anderson
4 Pender
5 Heth
6-9 Ewell (2nd Corps)
6 Rodes
7 Gordon
8 Early
9 Johnson

Buford's fighting withdrawal bought time for Major-General John Reynolds's 1st Corps to come up from the south.

Reynolds deployed his men to meet the Confederate advance down the Chambersburg Pike. Initially the brigades of Meredith and Cutler opposed the advance of Archer's and Davis's Confederate brigades. For a while the fighting went against the Confederates. Archer's brigade was outflanked by Meredith's men and cut up badly. Archer himself was captured. Two of Davis's regiments were taken nearly intact in a railroad cut, and Davis's brigade was withdrawn from the fighting. Pender's division was coming up, and these men stabilized Heth's shaky Confederate line.

Reinforcements were also arriving on the Union side. Reynolds had been killed by a sharpshooter in the morning's action, but most of his corps was up now, and these men would shortly be joined by Major-General Oliver Otis Howard's 11th Corps. (Command of the 1st Corps was taken over by Major-General Abner Doubleday.)

Howard's corps was deployed to the north and north-east of the embattled line of the 1st Corps. Two Confederate divisions were approaching from these directions, and soon Major-General Jubal Early's division of Ewell's Confederate 2nd Corps was overlapping the right flank of the 11th Corps.

Howard's 11th Corps, composed mostly of German immigrants, many of whom could not speak English, was held in contempt by the men of the Army of the Potomac. These 'Dutchmen', as they were styled, had panicked at Chancellorsville when hit in the flank by 'Stonewall' Jackson's sledgehammer attack. The army had unjustly blamed them for the loss of that battle. Now they were to be trounced again – this time by Gordon's brigade of Early's division.

Brigadier-General John Brown Gordon, the man who could 'put fight into a whipped chicken', determinedly drove his men into Brigadier-General Barlow's division of the 11th Corps. After a brief but savage resistance, the Union line began to break, and a rout developed with jubilant Confederates following the beaten 11th Corps into Gettysburg.

Now the stolid 1st Corps began to feel the pressure of the massive, renewed Confederate attack to its front. From Rodes on their right to Pender in their front, the whole Confederate line seemed to heave forward with desperate energy. With the 11th Corps fleeing into the town behind them, there seemed to be but one option left to the weary 1st Corps. They must withdraw fighting, and perhaps by their sacrifice gain time for the rest of the army to arrive.

The Confederate advance was irresistible. Perhaps only the timely appearance of Buford's cavalrymen on the left saved the 1st Corps from the fate of the 11th. Buford bluffed a charge, and the nearest Confederate units formed square – perhaps the only instance of this formation being employed during the entire Civil War. This change in formation broke the impetus of the Confederate attack and allowed the 1st Corps to re-form on Cemetery Hill behind the town.

General Lee arrived on the field at 3 p.m. Within the hour he had witnessed the rout of two Union corps. Lee's men had captured nearly 6,000 prisoners and several abandoned pieces of artillery. The two Union corps had been very nearly wrecked but had managed to establish a defensible line directly behind Gettysburg on Cemetery Hill.

On this low hill urgent attempts to rally the exhausted Union troops involved in the actions of the morning and early afternoon were being made by Major-General Winfield S. Hancock. By his own admission, Hancock 'had not been very successful'. This was the ideal moment for Lee to strike at Cemetery Hill.

But Lee was hesitant. He had no idea what force the enemy had at hand, and Longstreet's 1st Corps had not yet arrived from Cashtown. Still, the heights were the key to the field, and Ewell was instructed to 'carry the hill occupied by the enemy, if he found it practicable, but to avoid a general engagement until the arrival of the other divisions of the army'.

In other circumstances Ewell might have pressed the attack savagely, but now a strange, unexplained caution possessed him. Gordon, whose troops were already clambering up the wooded slope of Culp's Hill, which flanked and commanded the Union position, was ordered to halt and withdraw. Major-General Isaac Trimble, a hard-bitten veteran temporarily without a command, stormed at Ewell, but the commander of the Confederate 2nd Corps, overcome by 'physical fatigue and mental exhaustion', as Colonel Walter Taylor later put it, declined to push his men further. As the afternoon shadows lengthened on the field this extraordinary opportunity slipped away. Amid the declining sound of musketry Confederates in the town began to discern the steady noise of picks and shovels at work on the heights.

All through the late afternoon and evening of 1 July the Union army toiled toward Gettysburg in forced marches that left thousands of men prostrate with exhaustion. Meade had ordered his march in such a manner that his seven army corps were in close supporting distance of one another, yet still covering Washington and Baltimore. By chance Gettysburg was the nexus of twelve major arterial roads, and Meade was taking advantage of their good marching surfaces rapidly to concentrate his army.

The Second Day, 2 July 1863

By 2 a.m. Meade was on Cemetery Hill assessing the situation. The position occupied by the Union army was defensively strong. The 3rd Corps of General Sickles had taken position to the south and left of the line on Cemetery Hill. Hancock's strong 2nd Corps was within easy supporting distance to the south and would arrive at daybreak. Slocum's 12th Corps would shortly occupy the strategic wooded height of Culp's Hill, extending the Union right. The 5th Corps was bivouacked a few hours away; likewise Sedgwick's 6th Corps, which was expected in the afternoon.

From right to left the Union line occupied high ground. The line resembled a giant fish hook with the twin anchors of Little Round Top to the south and Culp's Hill to the north-east. Each part of the line was within easy supporting distance of others, and the nature of the position offered an unparalleled opportunity for the skilled Union artillerymen to hit their targets. After conferring with his subordinates, Meade decided to stand and fight. He was, according to his son, 'in excellent spirits, as if well pleased with affairs as far as they had proceeded'.

At 5 a.m. on Seminary Ridge, opposite the Union line, Lee was in a similar conference. The Confederate plan, as it evolved from that conference, was for Longstreet to take his fresh divisions of Hood and McLaws on a concealed march to the south of the Union left flank, supposed to be resting at that time on the Emmitsburg Road. The Confederate divisions would turn and face to the north-east, forming a line nearly at right-angles with the Emmitsburg Road and the Union left. Longstreet's attack would then be made on the exposed flank of the Union line. Part of A. P. Hill's corps was to co-operate by attacking Meade's left-centre (Cemetery Ridge). Simultaneously, Ewell's men were to strike at the Union right.

An attack of that nature requires perfect co-ordination and flawless execution. This attack had also to be made before the disparate elements of Meade's army were united. But Longstreet's corps, at that time strung out for miles over the rolling farmland behind Seminary Ridge, was not in a position to open the attack until 3 p.m. A study of the records shows that Longstreet might possibly have been ready by 9 a.m. had he been ordered to take up position. But Lee gave no specific order until 11 a.m. This situation was further aggravated when delays in Longstreet's march seemed to add substance to the allegation that he was 'sulking' because Lee had not approved his pet project of interposing the Confederate army between Meade and the city of Washington.

The arguments about what happened that morning will probably never be resolved, but recent research has shown that even had Longstreet been in position to strike at 9 a.m., his attack would probably have failed. The Union position at 9 a.m. was actually *stronger* than it was at 3 p.m. In those six hours General Sickles, commanding Meade's 3rd Corps, obligingly moved his men from the ridge line to the Emmitsburg Road. Sickles's forward movement exposed his corps to Longstreet's attack and left it in a position where it could not be very easily supported. It also completely denuded the Round Tops of troops.

A Union signal station on Little Round Top had reported Confederate movement away from the Union left and towards Emmitsburg. This was Hood's division of Longstreet's corps marching into position for the attack. On being informed of this movement Meade despatched General G. K. Warren to Little Round Top to investigate, and at 3.30 p.m. went himself to Sickles's field headquarters, probably with the intention of correcting Sickles's unauthorized dispositions. Sickles sought to defend his new position by pointing out to Meade that he now held higher ground, to which Meade replied, 'General Sickles, this is in some respects higher ground than that to the rear; but there is still higher ground in front of you, and if you keep on advancing you will find constantly higher ground all the way to the mountains.'

Sickles then declared that he would move his corps back to the line of Cemetery Ridge, but at that moment, Longstreet's attack struck. Thoroughly annoyed, Meade snapped, 'I wish to God you could, Sir, but you see those people do not intend to let you.'

Meade had not expected an attack on this front, but seeing that a serious Confederate drive had opened there, he began to direct reinforcements to Sickles. The 5th Corps and most of the 12th Corps were to be sent, but Sickles must hold his position until they arrived.

For two hours Sickles's men held the angle of their salient at the Peach Orchard on the Emmitsburg Road. McLaws's division attacked them in front and from both flanks, while Hood's division worked towards the Round Tops against negligible opposition.

Some of John Bell Hood's Texas Scouts had brought him news that the Round Tops were unoccupied, and against Longstreet's specific order to carry out the attack as planned, he siphoned most of his men away from the stand-up fight near the Peach Orchard and sent them towards the Round Tops. Hood's men were from the frontier South, and they must have felt at home as they entered this new area of woods, tangled scrub and huge boulders – known, incidentally, to the local Pennsylvania farmers as Devil's Den.

As Hood's men made their way forward, miraculously maintaining formation, they were startled by a single shell fired into their ranks from a mass of boulders to their front. This shell had been ordered by General Warren on Little Round Top, and the sudden start in Hood's ranks had sent a reflection of sunlight gleaming off a thousand musket barrels and revealed to Warren that Hood's men were behind Sickles and ready to ascend the Round Rops (indeed some were behind the Round Tops). This was the crisis of the battle, and what Warren did next earned him the title of 'Saviour of the Union'.

While his signal officers waved their flags wildly to deceive the enemy, Warren sent for troops. Sickles could spare none. He was outnumbered, and his Peach Orchard salient was caving in. Vincent's brigade of the 5th Corps, advancing to support Sickles, was detached and hurried over to the valley between the Round Tops. Warren personally led the 140th New York Regiment and Hazlett's battery of Weed's brigade to the crest of Little Round Top. His men began to arrive as the Confederates started their ascent.

Hood's men came on, stumbling and clambering up the difficult slope. As they advanced they gave out the high-pitched 'Rebel yell', which was distinctly audible to the Yankees similarly engaged in climbing the reverse slope. The Yankees reached the crest first, and a desperate struggle ensued at close quarters with neither side willing to give ground. Vincent was forced back in the valley between the Round Tops, but on Little Round Top the 20th Maine Regiment, having fired all its cartridges in less than ten minutes' action, rushed the Confederates in a wild charge. The Confederates were dispersed behind rocks and boulders, and although they outnumbered the 20th Maine, they were not formed and were successfully driven off.

During these savage exchanges Sickles was attempting to extricate the remnants of his beleagured corps from the Peach Orchard. As his men drew back, pursued by jubilant

left A general view of the battle during the second day, as seen from Little Round Top looking along Cemetery Ridge, which was held by Meade's troops (right). **below** Pickett's ill-fated charge on the third day is shown from a viewpoint behind the Union lines.

Confederates from McLaws's division, Sickles had his leg smashed by a bullet. Hancock took command of the 3rd Corps, and with men of the 5th and 2nd Corps established a line behind the Peach Orchard at the edge of a large wheat-field. Further to the right, heroic and, at times, unsupported resistance by Union batteries held up the Confederate advance.

The fighting in the wheatfield defies description. Men have referred to it as an 'inferno' or a 'maelstrom' and Lieutenant Frank Haskell, watching from Cemetery Ridge, found the air filled with a 'mixture of hideous sounds'. Men stood upright forty yards apart and for nearly two hours exchanged the most destructive fire they were capable of producing. Finally, the Union line retreated across Plum Run where it was reinforced by the 6th Corps.

Further north, Barksdale's Mississippi brigade advanced towards the Union line on Cemetery Ridge. The unfortunate Barksdale had chosen this day to go into battle in full Masonic regalia. By chance, his brigade charged a part of the line held by several New York Irish regiments. Barksdale was a marked man, and his body fell riddled with bullets.

Still further north, Anderson's division of A. P. Hill's corps joined the progressive but unco-ordinated advance of the Confederate right-centre. Two of Anderson's brigades, Wright's and Posey's, actually broke into the Union line after crossing three-quarters of a mile of undulating open ground under heavy fire. Wright's men took a line of guns but were not supported. A savage charge by the Union 1st Minnesota Regiment, which lost eighty-two per cent of its men, drove Wright back, but his brief success against the Union centre was to have ominous implications for a Virginia division commanded by Major-General George Pickett.

On the Union right, Ewell's Confederate 2nd Corps began its advance as Longstreet's attack petered out. Edward Johnson's division took some earthworks on Culp's Hill abandoned by Geary's division of the Union 12th Corps when it was sent to Sickles's aid. But Johnson's men failed to follow up their success and were driven out when Geary returned. Jubal Early's division of Ewell's Confederates made a spirited dash up Cemetery Hill and captured a battery, but was raked by guns aimed at its flank. The Union 11th Corps redeemed its reputation by refusing to move from its position behind the guns Early had overrun, and when Rodes would not advance from the west in support, Early was forced to withdraw. The last shots of the second day's action were probably exchanged as Early's men trickled back the way they had come.

The Third Day, 3 July 1863

Sometime during the evening of 2 July Meade assembled his corps commanders to consider a plan of action for the next day. The Union army had been hit hard (it was later estimated that sixty-eight per cent of all the casualties sustained by Meade's army at Gettysburg were inflicted on 2 July) but all the senior officers were in favour of staying to resist a Confederate attack. As the council adjourned, Meade turned to Brigadier-General Gibbon, who commanded a division of the 2nd Corps posted on Cemetery Ridge, and said, 'If Lee attacks tomorrow it will be in your front.' Startled by this statement, Gibbon asked Meade why, and he replied, 'Because he has made attacks on both our flanks and failed, and if he concludes to try it again it will be on our centre.'

Meade's statement was prophetic, but that evening Lee was considering not an attack on the Union centre, but a resumption of the general plan governing the attacks of the second day. He felt that 'proper concert of action' had been lacking, and that if the attacks could only be better co-ordinated and reinforced with fresh troops, a great victory might still be gained.

Circumstances, however, would not allow this plan to be implemented again. For the first time in the course of the battle Meade's army took the initiative and, in doing so, wrecked one of Lee's pincers before it could renew its assault.

By 10.40 a.m. on 3 July Edward Johnson's Confederate division of Ewell's corps had been knocked out of the Union entrenchments on Culp's Hill in a hard fight that had gone on since dawn. Longstreet had been directed to renew the Confederate attack on the Round Tops, but he was personally opposed to any further direct assaults on this flank and delayed his advance. Sensing failure on both flanks, Lee decided to stake everything on a grand assault by three divisions on the Union centre. This was a desperate decision – more so in retrospect – but Lee had overwhelming confidence in the ability of his men to break the Federal line.

Longstreet bitterly opposed this plan, stating that, in his opinion, 'no 15,000 men ever arrayed for battle can take that position'. But Lee was firm. Longstreet was to direct the attack, which was to be made by Pickett's fresh division of his own corps with, from A. P. Hill's corps, Heth's division (now commanded by Pettigrew) and two brigades of Pender's division under the command of Trimble.

The assault was to be preceded by a massive artillery bombardment meant to silence the numerous Union artillery on Cemetery Ridge and pave the way for the infantry. Colonel E. P. Alexander of Longstreet's corps artillery was to supervise the fire of the guns and give the signal for the advance when he judged the Union batteries to have been silenced or withdrawn.

Although there was enthusiasm among many Confederate officers at the prospect of the assault – Pickett was reportedly 'congratulating himself on the opportunity' – the severe problems it posed weighed heavily on more rational minds. Firstly, it seemed improbable that the Confederate artillery could be nearly as effective as was hoped, no matter how concentrated its fire. The guns closest to the Union line were firing from three-quarters of a mile, the farthest from over two miles; precise shooting was impossible. Secondly, some of the troops making up the assaulting column had been demoralized in the earlier fighting and might go to pieces under fire. Thirdly, no adequate provision was made to support the flanks of the columns with either infantry or artillery. And finally, the changed conditions of warfare had made such a magnificent effort, truly in the manner of Frederick the Great or Napoleon, virtually impossible. For what was envisioned was not so much the storybook image of massed ranks of infantry charging directly at the Union centre, but rather two assault columns, each with a front of roughly half a mile, emerging from the woods of Seminary Ridge at points about one thousand feet apart, and then trying to storm the enemy.

The map shows the dispositions of the rival armies at about 2.30 p.m. on the third day, half an hour before Lee flung the greater part of three divisions at the Union centre in a last and desperate effort to break the enemy's line. It was an ill-judged venture; the assault columns of Pickett – on the right – and the joint force of Pettigrew and Trimble on the left advanced bravely but were mercilessly smashed by the combined fire of Meade's infantry and guns.

UNION

Army Corps
Cavalry Corps
1 Merritt (Cavalry Corps)
2 Kilpatrick (Cavalry Corps)
3 Sedgwick (6th Corps)
4 Sykes (5th Corps)
5 Sickles (3rd Corps)
6-8 Doubleday (1st Corps)
6 Doubleday
7 Robinson
8 Wadsworth
9 Hancock (2nd Corps)
10 Howard (11th Corps)
11 Slocum (12th Corps)

CONFEDERATE

Army Corps
1-3 Longstreet (1st Corps)
1 Hood
2 McLaws
3 Pickett
4-7 A P Hill (3rd Corps)
4 Anderson
5 Heth (now commanded by Pettigrew)
6 Pender
7 Trimble (Pender)
8-10 Ewell (2nd Corps)
8 Rodes
9 Early
10 Johnson

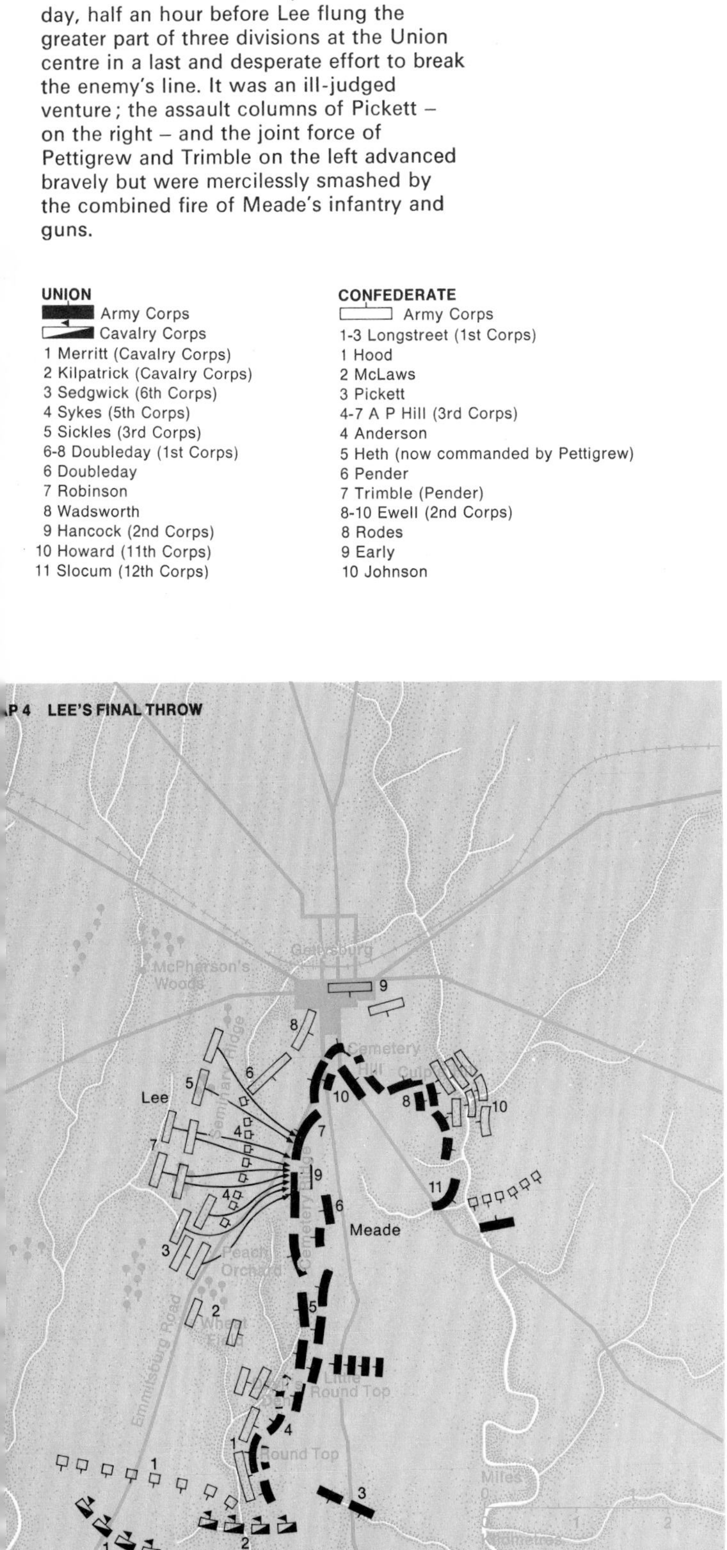

At 1.30 p.m. two signal shots were fired from a Confederate battery in the Peach Orchard, and within a few seconds, according to a Union observer on Cemetery Ridge, 'the report of gun after gun in rapid succession smote our ears and their shells plunged down and exploded around us'.

At 2.55 p.m. the artillery bombardment abruptly ended, and five minutes later the Confederate infantry began to move forward. Their advance was so perfectly executed as to arouse feelings of awe and admiration among their enemies. In Lieutenant Frank Haskell's eloquent words, the men approaching so deliberately moved 'as with one soul, in perfect order . . . magnificent, grim, irresistible'.

But soon the Union artillery began to smash great gaps in Pickett's lines. First the long-range fire of shell and solid shot – the guns on Little Round Top discharging a murderous enfilade fire. Further on, the Confederate brigades met the big shot-gun blasts of canister, later combined with sweeping volleys from the infantry. On the left of the charge, Brockenbrough's Virginia brigade broke up under the fire. But a great wedge of troops still advanced.

Finally, Brigadier-General Armistead led perhaps three hundred of Pickett's and Pettigrew's men into the Angle, the objective of the charge. Armistead was killed immediately, and the pitifully few men who followed him were swept away by Union reinforcements. This was the end of the battle and, symbolically, the turning-point for the Confederacy.

Aftermath and Conclusion

The Union army lost nearly 22,000 men in the battle and Confederate losses have been put as high as 28,000 and as low as 20,500. The losses among officers for both armies were especially high and some authoritative sources state that Lee's army never recovered from this deficiency.

The fortunes of the Confederacy were shattered at Gettysburg, and what followed was a grim, tragic and needless retreat that lasted for two long years. The Army of Northern Virginia was never again the confident and magnificent offensive weapon it had been before Gettysburg, but modern weaponry and good leadership combined with an unwavering, primitive faith in the righteousness of its cause produced a finely-honed defensive machine which held the Army of the Potomac at bay until the Confederate surrender at Appomattox on 9 April 1865. This signified the end of the American Civil War; at a total cost to the nation of over 600,000 lives, peace was achieved and the first steps taken towards emancipation.

Donald Featherstone at

EL ALAMEIN

In the Second World War the Western Desert campaign stands out as a remarkable episode – a trial of strength between the tanks and infantry of the Allies and those of Rommel's Axis forces. The climax of this strangely isolated war, grimly waged in alien surroundings, was a twelve-day battle at El Alamein, where General Montgomery ended by breaking the Axis hold on North Africa.

1942

On 23 June 1942 Rommel re-entered Egypt; this illustration shows German tanks advancing along the coast road to Sollum.

Background to the Desert War

From a strategic point of view, the Western Desert was not the most important front in the Allied struggle against the Axis during the Second World War. Both sides, however, led respectively by the British and the Germans, needed to dominate the Mediterranean, and once Italy had entered the war on the Axis side in 1940 it was inevitable that there would be a confrontation in the desert. This would happen when Italian forces from Libya (then an Italian colony) attacked the western flank of the British position in the Middle East. For the British this was a crucial region, from which their forces could provide support to Turkey and also, in Egypt, operate a half-way base on the route to India. Furthermore, had the Nile Delta been available to the Axis forces, it would undoubtedly have served them as a springboard against Russia's southern borders, in which case the danger to the Soviet Union would have been extreme – at a time when the latter was engaged on the Eastern Front in the largest and most decisive battle of the whole of the Second World War. But, perhaps most of all, the Desert War was a battle for oil: the Middle East, quite simply, contained the oil fields on which the Allied war effort depended.

In the Western Desert it was not territory that was sought. Both sides realized that there was little virtue in capturing great tracts of sand. The more desert that was conquered, the more supply problems were aggravated and the more vulnerable communications became. The aim of both sides was to locate and destroy the enemy. On the Axis side the German Afrika Korps worked on the principle that once the Allied 8th Army was thoroughly defeated then the Nile Delta would be laid open to them. In this they reckoned without the El Alamein position or the determination of General Montgomery. The vital geographical factor was that there were no defensive positions in the desert that could not be outflanked – except at El Alamein and at El Agheila. At these two places secure flanks were provided by narrow bottlenecks between the Mediterranean on the one side and, on the other, by the Qattara Depression at El Alamein, and by the Great Sand Sea at El Agheila. Everywhere else an open desert flank caused constant anxiety to all army commanders.

In the Desert War victory and defeat hinged principally on supplies: mobile forces could advance or retreat only so long as supply difficulties did not bring them to a halt. Everything – water, rations, ammunition, POL (petrol, oil and lubricants) as well as spares had to be transported from the main bases of Alexandria and the Suez Canal district for the British and from Tripoli for the Germans. The scattered and thinly distributed supply bases set up in the desert by both sides were vulnerable to aircraft, to long-range raids and to the rapid advances of the war itself.

These considerations played a big part in the eventual defeat of the Afrika Korps because throughout the later

The Desert War began in June 1940, when British forces from Egypt attacked Italian bases in Libya, and lasted until the Axis surrender of 12 May 1943. In April 1941 Rommel pushed the British back into Egypt – except for a garrison at Tobruk. In November Tobruk was relieved and Rommel was driven west beyond El Agheila. Three weeks later newly reinforced, Rommel began an advance that took him rapidly back to Tmimi and Gazala; after a lull, Tobruk fell to him in June 1942 and Axis tanks swept far into Egypt. The Allies withdrew to the El Alamein position; there, deployed along a 40-mile front, their flanks protected by the Mediterranean and the 700-foot cliffs of the Qattara Depression, the Allies prepared to bring an arduous campaign to its climax.

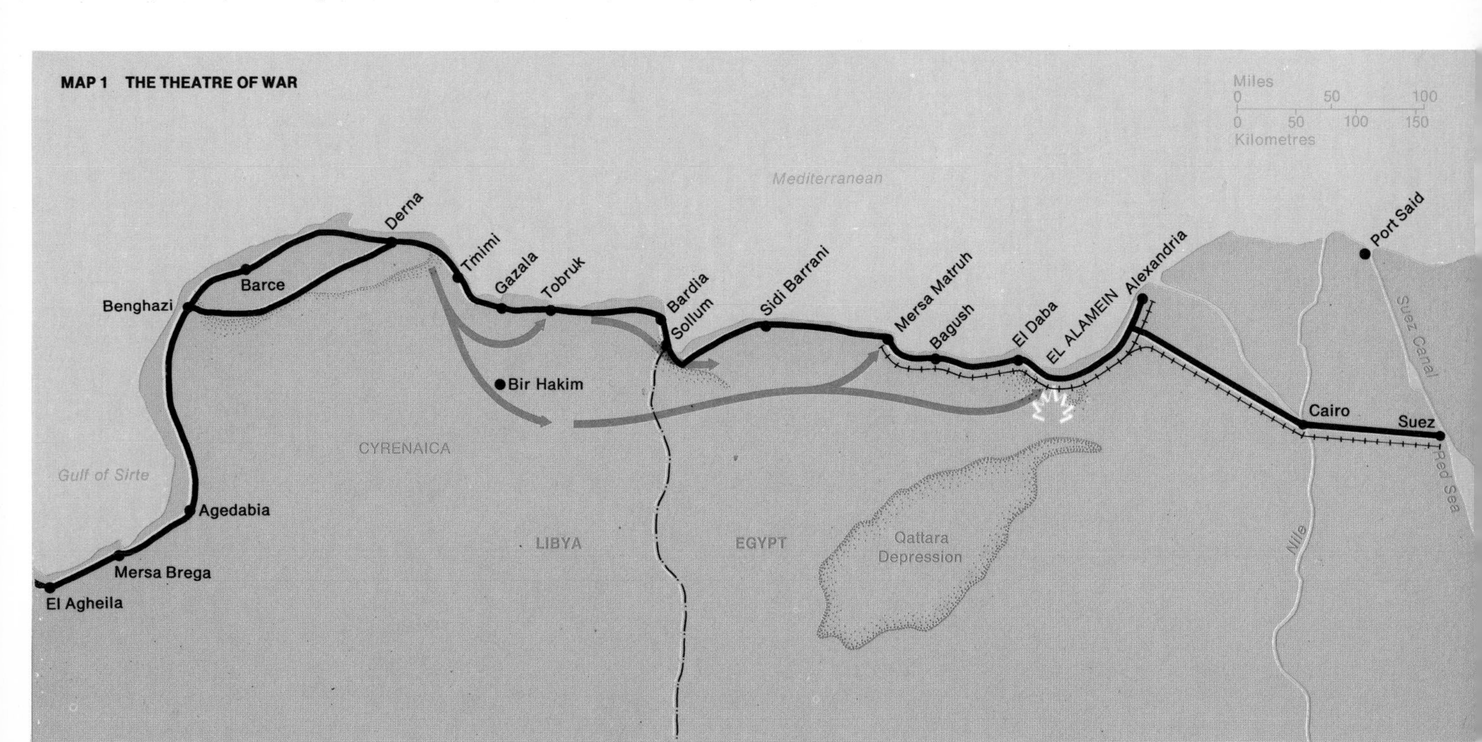

stages of the Desert Campaign Rommel's bases, supply lines and supply ships were constantly attacked by the Royal Air Force and the Fleet Air Arm. At the time of El Alamein the Royal Air Force and the Army were co-ordinated in an unprecedented manner in North Africa, where it was a matter of policy for 8th Army and Desert Air Force Headquarters to be set up alongside each other.

Despite the difficulties caused by lack of water, the sense of emptiness and loneliness exaggerated by discomforts and disappointments, men rather than weapons were the decisive force in the desert. Allied soldiers, the British, Australians, New Zealanders, South Africans, Indians, Poles, Czechs, Greeks and French, took well to the desert and went along with it. Bronzed and tough, they were one of the fittest armies in history. The Germans were good desert soldiers: they moved fast and were well fed, they had fine equipment and were brilliant at map reading and navigation. Nevertheless, they loathed the desert and it was unkind to them. The Italians regarded the desert romantically rather than practically and never came to grips with it in their attempts to turn this inhospitable alien terrain into a land of their own image. The best desert soldier realized that the desert was something to be used and not feared.

War in the desert could be likened in many ways to war at sea, the reconnoitring armoured cars taking the destroyers' parts, while the guns and tanks were the battleships and cruisers. Similarly, just as sea power in the Second World War could be destroyed if it was not adequately supported in the air, so were the desert armies open to destruction if one side had air superiority. There was a further resemblance to sea warfare in the manner in which movement was restricted by the minefields which covered vast, often uncharted areas.

It was soon obvious that the British principle of attaching tank forces to infantry units was inferior to the German concept of armoured tactics. Moreover, the German commanders in the Desert were superior to their British counterparts in that they made fewer mistakes and took correct decisions more often. Until Montgomery arrived, Rommel was an abler general than any on the British side and the army he commanded was a more professional force than the British Army. What Rommel had done was to elaborate the basic German battle drill to achieve a balance of mobile striking power, using tanks, anti-tank guns and field artillery together as a common force.

The German control of armour was superb because they were a highly trained group of self-contained tank technicians controlled *en masse* as easily as a single vehicle. In action, tanks, anti-tank guns, recovery vehicles and petrol wagons, supported in the air by dive-bombers, went forward rapidly and successfully, often accompanied by senior officers. The co-operation between the armour and anti-tank gunners was exceptional, in fact the brilliant successes of the Afrika Korps depended upon three factors – the

superiority of their anti-tank guns, the systematic co-operation of all arms and their tactical methods. In the long run it was equipment and training that counted. The Germans fought as a team whereas the British, despite their brilliance and dogged spirit, fought as individuals until they learned better. They did not seem to be able to marshal and drive their tanks as the Germans did – perhaps they were simply not trained so highly as the much-practised Germans; perhaps, too, they did not possess the same innate feeling for armour.

When the Afrika Korps under Field Marshal Rommel arrived in the desert in 1941 they brought with them ideas and equipment that had been successfully used during the

left Men of Rommel's Afrika Korps (DAK).
below Desert troops and equipment:
left A German 88-mm gun captured by Australian troops. **centre** A British Scorpion flail tank used for minesweeping. The drum rotated rapidly, causing lengths of heavy chain to beat a tattoo on the ground and so explode mines in the Scorpion's path.
right An Allied field telephone in operation beside a knocked-out German tank.

right Axis forces in a typical desert formation during the outflanking of Bir Hakim, on the road to Egypt. Bir Hakim itself was evacuated in June 1942 after a ten-day defence.

blitzkrieg in Poland and France. They followed a system of over-all offensive that demanded close support between soldiers on the ground and aircraft, the dive bomber being used as airborne artillery. At the same time that he sent Rommel and his men to Africa, Hitler ordered Messerschmitt 110s and Stukas to go with them. The Stuka dive bomber had been a great success in France and Poland; in the Battle of Britain they had been cut to pieces by Spitfires and Hurricanes – but in the desert there were no Spitfires and few Hurricanes. In 1940 the only fighter aircraft available to the British in the Middle East were some Gloster Gladiator biplanes, antiquated and outmoded aircraft that stood little or no chance against the German fighters. As the campaign progressed, however, the British gradually gained in air power until, at the time of Montgomery's arrival in the desert in August 1942, they were masters of the sky.

In June 1940, British forces crossed the line into Libya and shelled the forts of Capuzzo and Maddalena. These were the first steps in a long and hard Desert War. For fifteen months following that first offensive the Allied and Axis forces pushed each other up and down the desert.

Chronology 1

Sept 1940–Feb 1941
Italians under Graziani advance towards Egypt. General O'Connor launches Allied Western Desert Force and in two months has smashed Italian Army and captured Cyrenaica.

31 March–mid-April 1941
Rommel arrives and sends his Light Division forward. Pushes British back behind Egyptian frontier, except for garrison at Tobruk. General O'Connor captured.

15 June 1941
Operation 'Battleaxe' fails with heavy losses to recently arrived Crusader tanks. General Wavell relieved of Middle East Command and replaced by General Auchinleck.

18 Nov 1941
Operation 'Crusader'. Tobruk relieved. Rommel pushed back, first to Gazala, out of Cyrenaica to the position beyond El Agheila from which he had started nine months before. Three weeks later Rommel, newly reinforced, chases British back to Tmimi and Gazala.

Field Marshal Erwin Rommel
Axis Commander-in-chief
Field Marshal Erwin Rommel (1891–1944) was a 'soldier's general'. He possessed all the dash and élan of a great cavalry leader – and indeed was not unlike J. E. B. Stuart from the American Civil War. At the time of El Alamein, Rommel was in general physically tougher and lived harder than most of the men he commanded – this despite an illness which kept him from the field during the first two days of the battle and which obviously proved helpful to Montgomery in the early stages. Under normal conditions Rommel's initiative led him to the heart of the action where his dash, courage and electric presence raised the morale of all around him. Intuitively grasping new tactical and technical possibilities, Rommel was adept at learning from the experiences of others, and an essential element of his victories in Africa was his talent for converting mistakes into successes. He was prepared to take tactical and administrative risks, and in a lesser commander this aggressive spirit could well have been dampened by the tight control exercised on him by his political masters, who tended to regard the African campaign as a sideshow. Throughout the war, and particularly in the Western Desert, Rommel never failed to be held in honour by his opponents.

Lt-General Bernard Montgomery
Allied Commander-in-chief
Lieutenant-General Bernard L. Montgomery (b. 1887) was one of the few British generals in the Second World War who could match the expert and self-assured professional commanders turned out by the German General Staff system. As Commander-in-chief of the British Armies, first in Africa and then in north-western Europe, his essential role was to achieve success with the steadily waning manpower of a comparatively small nation. This made it necessary for him to refuse to involve his men in any operations where the outcome was less than reliably certain. His use of armour at El Alamein and in the chase afterwards provide a strong case for asserting that he could have pushed on more rapidly and to greater effect. But if Montgomery was no great handler of tanks he could handle men and not only stimulated his immediate subordinates, but also made the man in the ranks feel that he was an essential part of a first-class army. At the same time he sometimes lacked the ability to communicate his intentions to his equals and superiors. Nevertheless, for the past 150 years there has been no man in the British Army to surpass Montgomery for sheer professionalism and sustained success in the field.

There followed a lull of some four months during which both sides milled around and tested each other in the dust bowl between Tmimi and Gazala. The 8th Army had set up a series of defensive positions, the Gazala Line, covered by extensive minefields around large strongpoints sited for all-round defence.

The War Reaches El Alamein

General Ritchie, the 8th Army Commander, had disposed his forces on the assumption that Rommel would attack in the north. At the same time he was preparing his own offensive for early June as ordered by Auchinleck, under pressure from Winston Churchill, the British Prime Minister. Rommel beat them to the punch.

Chronology 2

26 May–mid-June 1942
Rommel outflanks Ritchie in south and pushes back Allies. Bir Hakim evacuated by Free French after ten-day defence. Rommel embarks on second siege of Tobruk. Tobruk falls and garrison of 25,000 surrenders.

23 June 1942
Rommel re-enters Egypt.

25 June–1 July 1942
General Ritchie relieved of 8th Army command by General Auchinleck. 8th Army rallies and German advance halted at El Alamein.

At this stage, the Afrika Korps was as weary as the 8th Army and Rommel was unable to break through the Allied lines. The Germans had by then outstripped their supplies of petrol and ammunition and all their vehicles were badly in need of maintenance.

To understand fully the vital struggle that followed, it is necessary to know something of the El Alamein position.

Between the Mediterranean and the 700-foot cliffs that line the northern edge of the Qattara Depression lay a forty-mile stretch of desert. Its northern half was so featureless that almost imperceptible rises in the ground became of great tactical importance. To the south of the two most prominent of these ridges, Miteiriya and Ruweisat, were abrupt escarpments that made progress in the direction of the Qattara Depression increasingly difficult.

Just as Wellington had noted the Waterloo position before he fought on it, so had Auchinleck pre-selected this area as a suitable position for the final defence of Egypt if ever the situation demanded it. He made this choice for two reasons.

Firstly, if the forty-mile wide neck were held in adequate numbers (two full-strength infantry divisions and a large armoured force) it could not be turned; secondly, the unsuitable ground in the south would very much cramp the movement of armour.

All through July there was fighting but neither side was able to gather enough strength for a decisive blow. Massing his artillery and carefully nursing his armour, Auchinleck had reorganized the 8th Army into battle groups consisting of infantry and artillery in self-supporting brigades. During this period, the Germans, at the end of their supply lines and harassed by the RAF, were forced to end their attempts to break through and instead to concentrate on warding off Allied counter attacks. A German attack on 2 July was beaten off by British tanks strongly posted on Ruweisat Ridge and the German 90th Light was unable to break through the South African positions. Auchinleck was beating the Afrika Korps at its own game by drawing the Panzers on to his own armour and artillery, which were strongly dug-in in carefully chosen positions. Another feature of Auchinleck's tactics was systematically to attack the Italian formations, conscious that the German troops on whom Rommel mainly relied were too few in number and too exhausted to be able to succeed alone. Overhead, the Allied Desert Air Force commanded the battlefield.

Auchinleck's tactics were ably demonstrated during the battle fought by the 5th Indian Brigade near the Deir el Shein Depression. During the afternoon of 16 July German artillery pounded the Brigade's position whilst Stukas screamed down on them in waves. Towards evening, when the sun was well down in the sky and shining into the eyes of the defenders, the German tanks came out of the west and north-west. For three hours they tried to break through but the 6-pounder anti-tank guns and tanks in the hull-down position fought them off. When the Germans retired the battlefield was strewn with burning wreckage – the Germans had lost twenty-four tanks, six armoured cars and twenty-five guns.

The Battle of Alam Halfa

General Auchinleck was then replaced in the Middle East Command by General Alexander on 13 August and General Montgomery took command of the 8th Army. Seventeen days later Rommel struck again, but Montgomery had been given time to work out a course of action. He resolved to stand firm, forcing Rommel to come to him. 'We would fight a static battle and my forces would not move; his tanks would come up against our tanks dug-in in hull-down position at the western edge of the Alam Halfa Ridge.'

In the week-long battle that followed, Rommel sent his German and Italian columns in a right hook, attempting to force a way behind Montgomery's main defensive position. Repulsed in his thrust at the Alam Halfa Ridge, and handicapped by a shortage of petrol through attacks by the RAF, Rommel gave up the assault. The Germans had lost about 3,000 men, thirty-eight tanks and thirty-three guns; the British lost fewer guns but more tanks and about 2,000 men.

Now, at the end of a long and vulnerable line of communications, the Afrika Korps found itself forced upon the defensive – a most unattractive situation for an army accustomed mainly to offensive manoeuvres. However, Montgomery refused to move until he was completely ready, biding his time while abundant reinforcements of tanks, guns, troops and all types of supplies poured in.

The Rival Armies

The comparative strengths of the two armies at the Battle of El Alamein were approximately as follows:

Allied 8th Army		**Axis Forces**	
Men	195,000	Men	100,000
Tanks	1,345[1]	Tanks	510
Guns	1,900	Guns	1,325

[1] Of this number some 200 tanks came up during the action.

The 8th Army was organized in three Corps, on the following lines:

10th Corps (Lumsden)	**30th Corps (Leese)**	**13th Corps (Horrocks)**
1st Armoured (*Briggs*)	51st (Highland) Infantry (*Wimberley*)	7th Armoured (*Harding*)
10th Armoured (*Gatehouse*)	2nd New Zealand (*Freyburg*)	44th Infantry (*Hughes*)
8th Armoured (*Gairdner*)	9th Australian (*Morshead*)	50th Infantry (*Nichols*)
	4th Indian (*Tucker*)	
	1st South African (*Pienaar*)	

Less than three weeks after taking command of the Allied 8th Army, Montgomery successfully resisted an Axis right hook, the aim of which was to force a route behind the main Allied position and so isolate the 8th Army from its supply routes. But Montgomery's tanks stood firm and denied Rommel his key objective – the Alam Halfa Ridge. The map shows the extreme points reached by the Axis forces before they withdrew.

Montgomery's formidable array of armour has been calculated at some 285 Sherman tanks, 246 Crusaders, 421 Stuarts (Honeys), 167 Valentines, 223 Grants, a small number of Matildas and three Churchills. The Allied commander could also call upon the fire power of some 1,000 field and medium guns, 800 of the new 6-pounder anti-tank guns and 100 105-mm self-propelled guns.

Against this force were ranged these below-strength Axis Divisions:

German Afrika Korps (DAK) (Von Thoma)	**10th Corps (Nedda)**	**20th Corps (Stephanis)**
15th Panzer (*Vaerst*)	Folgore (*Frattini*)	Ariete (*Arena*)
Ramcke (Parachute) Brigade (*Ramcke*)	Brescia (*Brunetti*)	Trieste (*La Ferla*)
21st Panzer (*Von Randow*)	Pavia (*Scattaglia*)	

21st Corps (Navarini)	**Army Reserves**
Trento (*Masina*)	90th Light (*Sponeck*)
Bologna (*Gloria*)	164th Light (*Lungershausen*)
	Littorio (*Bitossi*)

above Allied equipment, including a Scorpion flail tank on a transporter with Indian carriers and a white scout car containing members of the 51st Highland Division. **right** German Panzer equipment on the move. Artillery, infantry and supply vehicles are shown to the left of the picture with 105-mm howitzers.

The Axis tanks were also heavily outnumbered, consisting of some 38 Panzer Mark IVs, 173 Panzer Mark IIIs and 300 Italian tanks of doubtful value (commonly referred to as 'self-propelled coffins'). Reserves barely existed. The Axis artillery was made up of about 475 pieces of which 200 were German and the remainder Italian short-range field guns; there were also some 850 anti-tank guns.

Initially, Montgomery planned to destroy Rommel's armour and then deal at leisure with the remainder of the Axis forces in their by now heavily mined position, forty miles long and five miles deep. Then, a fortnight before the battle was due to start, he reversed his thinking. His new plan entailed holding off or containing the enemy armour while he carried out a methodical destruction of the infantry divisions holding the defensive system. He proposed to attack them from both flank and rear, so cutting them off from their supply line.

In the open desert concealment was immensely difficult and it was impossible for the 8th Army to conceal that it was preparing a major thrust. Nevertheless its camouflage experts managed to a remarkable extent to disguise from the Axis forces the full extent of the build-up of troops, weapons and stores. Primarily the enemy had to be made to think that the attack would follow the tactical pattern of earlier battles and would be built round an assault on the desert flank. To this end, false staging areas were assembled in the southern part of the British position with wood and canvas armour, guns, dumps and pipelines all realistically laid out.

Montgomery's master plan in fact consisted of a break-through in the north by his 30th Corps, who were to smash through the Axis defences and cut two lanes in the enemy's minefields to allow the armour of the 10th Corps to pass through and control the Axis supply routes. This would force the Panzer divisions to attack them in terms favourable to the 8th Army. The Allied 13th Corps would attack in the south and so prevent the Afrika Korps from sending reinforcements to the northern sector. It was hoped that light armoured forces would be able to penetrate the enemy defences and move on to El Daba. This first, break-in phase of Montgomery's plan was code-named 'Operation Lightfoot'.

New Zealand infantry engaged in the pursuit from El Alamein pass a broken-down Crusader tank.

The Course of the Battle

The assault went in during the night of 23 October 1942, preceded by a tremendous artillery barrage. Then the 30th Corps fought its way on to Miteiriya Ridge while the 13th Corps made progress in the southern sector, although it was delayed by dogged resistance and two thick belts of minefields. Defending themselves stubbornly, the Afrika Korps did not immediately react by counter-attacking, possibly because Rommel had been away on sick leave since 25 September and his deputy, General Stumme, died of a heart attack during the first hours of the assault. In that time the Allied infantry pressed slowly forward through a fog of dust against determined enemy strongpoints; all routes were pitted with anti-personnel 'S' mines – lethal canisters whose three small wire prongs barely showed above the ground – and larger quantities of Teller anti-tank mines, which the sappers toiled to lift and so create safe passageways for the waiting tanks of the armoured divisions.

On the night of 23 October, the 7th Armoured Division had penetrated the first minefields in the south but was brought to a halt. Because it was required for a later attack, Montgomery withdrew it and called off that part of the attack. Later, on the following night, the 10th Armoured Division commander was doubtful whether he could push forward as ordered; however, Montgomery insisted that he advance and before 8.00 a.m. one brigade was 2,000 yards west of the minefield area. The New Zealand Division also fought its way clear. Counter-attacks by the 15th Panzer and Littorio Divisions were repulsed with heavy losses. In the meantime the task of clearing the northern corridor was successfully completed by the 51st Highland and the 1st Armoured Divisions.

Rommel reappeared on 25 October and took over from General Ritter von Thoma, who had replaced Stumme. His presence was badly needed as morale was beginning to sag, losses were heavy, petrol and ammunition were short and constant bombardment from the air was having a marked effect. The Axis resistance was nevertheless powerful and sustained. The German 88-mm anti-tank guns, in particular, poured out a deadly continuous fire and were a dominating force in the field. By the 26th Montgomery accepted that his original break-in plan had been effectively halted, and Operation Lightfoot was accordingly stopped. On the evening of the 27th Rommel counter-attacked but was beaten back from Hill 28 (Kidney Ridge) by a strong combination of anti-tank guns, tanks and bombing.

Montgomery regrouped for what he hoped would be his final assault. On 29 October the Australian Division advanced from the northern flank of the Kidney Ridge salient and made for the coast. As Montgomery had hoped, Rommel counter-attacked with his reserves, among them the 90th Light Division, but the Australians held firm and so the scene was set for 'Operation Supercharge', Montgomery's final attack.

This operation began on the night of 1–2 November. It consisted of an infantry push west and north-west of Kidney Ridge by the 7th Armoured Division and the New Zealanders, brought in from another part of the front. At first, the attack was held by the German anti-tank gun screen but the Afrika Korps did not have enough tanks or guns to absorb heavy losses and finally on 4 November the British broke through.

The armoured car regiments went through as dawn was breaking and soon the armoured divisions were clean away into the open desert; they were now in country clear of minefields, where they could manoeuvre and operate against the enemy's retreating columns. However, the general pursuit was a very cautious and slow-moving affair

After his success at Alam Halfa Montgomery prepared a massive build-up of troops and equipment with which he could strike decisively at the exhausted and outnumbered Axis forces, who were now also facing an acute shortage of petrol and ammunition. Montgomery's master plan was for his 30th Corps to carve two lanes in the Axis minefields to the north, letting the tanks of the 10th Corps through to seize control of the enemy's supply routes. In the south an assault by the Allied 13th Corps was planned to prevent Axis reinforcements from being sent north. The attack began on 23 October.

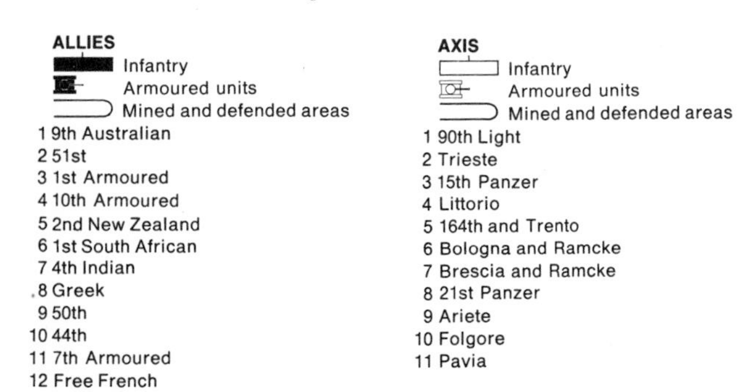

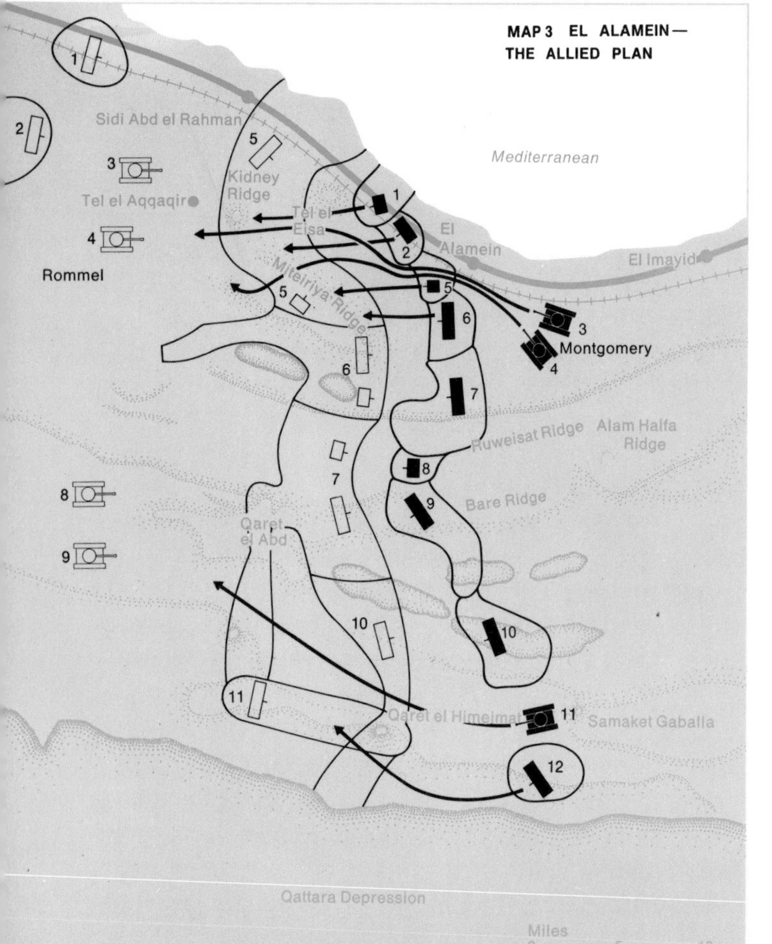

and was brought to a standstill by torrential rain on 6 November. Although they were overwhelmingly superior in armour and guns, the British at that point missed a great chance to destroy the Afrika Korps, but both victors and vanquished were exhausted by their twelve-day slogging match.

For his final thrust, due to begin on the night of 1 November, Montgomery planned an infantry push west and north-west of Kidney Ridge. At first the attack was held by Rommel's screen of anti-tank guns but a breakthrough was finally made on 4 November.

ALLIES
8th Army's front line
1 9th Australian
2 1st New Zealand
3 51st
4 1st South African
5 4th Indian
6 50th

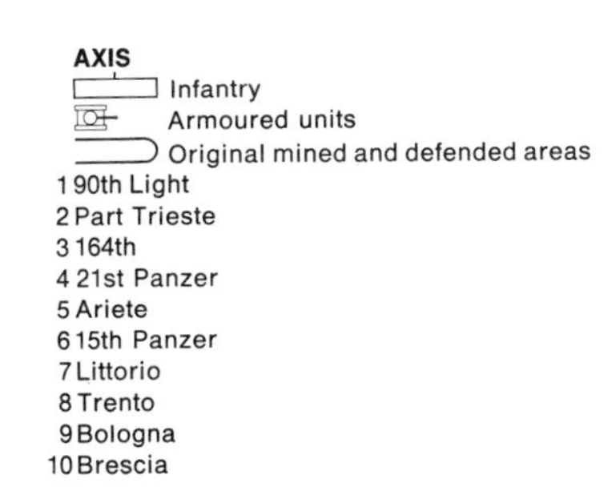

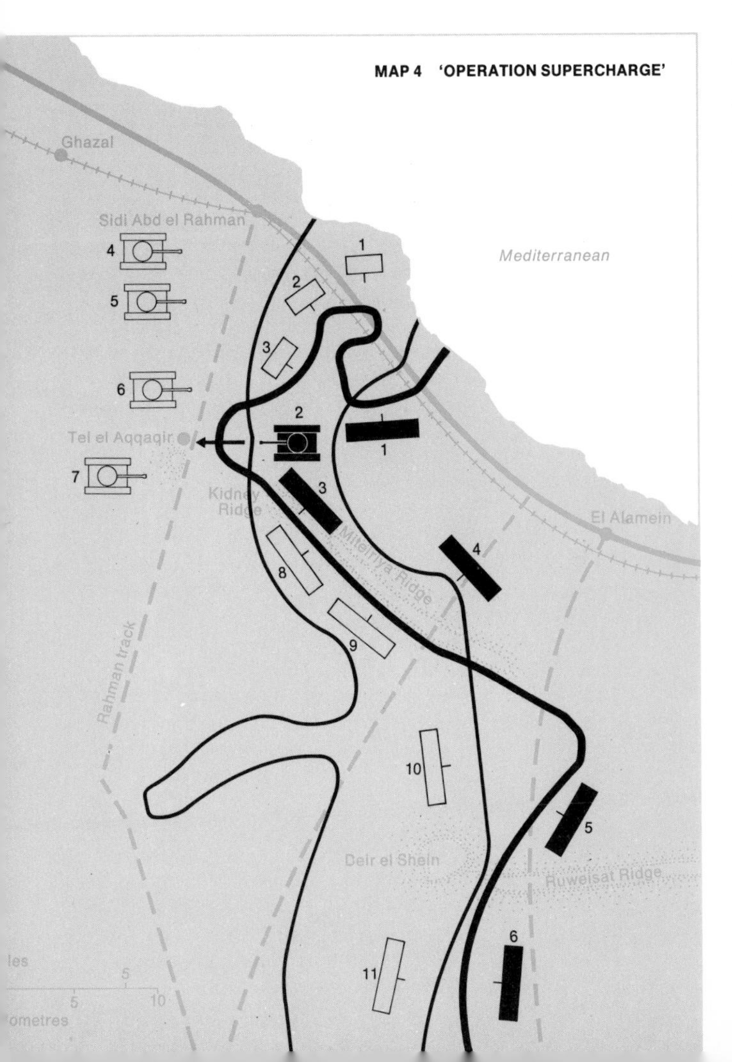

Aftermath and Conclusion

It is estimated that Rommel lost about half his force in killed, wounded and prisoners, together with 1,000 guns and 450 of his 500 tanks. By 15 November, he was considered to have no more than 80 tanks remaining. The British lost 13,500 men and had 500 tanks put out of action of which 150 were damaged beyond repair. About 100 guns had been destroyed. Without supplies and air cover, the Axis powers retreated and established a final foothold near Agedabia by the middle of November.

The Panzerarmee had not been destroyed but it had suffered such losses that never again could it take the offensive. The Allied pursuit was methodical and unrelenting but it was slow because after El Alamein the 8th Army was traversing Rommel's old stamping-grounds – El Adem, Knightsbridge, Sidi Rezegh. The scorched and rusted hulks that dotted those former battlefields aroused fearsome memories of the skill and rapidity of the Panzers attacking out of the setting sun. However, although the magic of Rommel still persisted, Tobruk was taken on 13 November, Gazala on the 14th and Benghazi on the 20th.

During the retreat, further engagements took place at Wadi Faregh, Buerat and the Mareth Line. In the meantime, strong American and British forces had landed in Morocco and Algeria on 8 November; Rommel now had to deal with an enemy on two fronts. At that point, Hitler decided to reinforce the deteriorating situation by throwing in more troops and armour in an attempt to stabilize the front and maintain his grip on North Africa. Even a 'Tiger' tank unit, the Panzer Battalion 501, was transferred to Tunisia. It was all in vain.

Towards the end of January 1943, General Eisenhower attacked with fifteen British and five American divisions. The end came on 12 May 1943, when approximately 150,000 German and Italian soldiers surrendered in Tunisia. The war in North Africa was over. As a result the underbelly of Europe lay open to any attack the Allies chose to mount. The final victory was mainly determined by the availability of supplies. The British had determined to fragment Axis communication lines in the Mediterranean and had done so. As a result the Axis presence in the Western Desert was quickly destroyed.

APPENDIX I

THE PRINCIPLES OF WAR-GAMING

The battles described in this book all belong to history. Many war-games are indeed based on the idea of re-fighting a famous battle such as Blenheim or Austerlitz – and a major purpose of the book is to set out for would-be war-gamers a series of realistic ground-plans which they may then use to build up their own war-games. However, not all war-games are of the 'straight' historical type, based on actual battles. Both for the sake of variety and because they like to test their ingenuity, many war-gamers prefer to devise their own hypothetical engagements which are staged under the conditions of a particular historical period.

In a war-game models or special pieces representing military units are manoeuvred into combat situations. These are then resolved in terms of the tactics and results of real-life conflicts. The pieces used are usually accurate scale-models of the soldiers and equipment they represent. The speeds of movement of the pieces, and the ranges of their missiles are scaled down from known data. To calculate the casualties from missile fire and from hand-to-hand combat, many war-gamers use specially devised tables. Other tables may be used to supply the human element of morale or 'fighting spirit', which determines whether a force will continue to fight, surrender or flee the field. Dice, too, are often used as a means of deciding those unpredictable factors which are a feature of all real-life conflicts, and which many war-gamers like to incorporate in their table-top battles.

To play a war-game, the following equipment is required – models of terrain, figures and equipment; dice; steel tape measures; pencils and paper; a set of rules and table space to play on.

Terrain

Terrain may be as carefully modelled and detailed, or as simple and symbolic as players prefer. The plates in this book show examples of terrain which is modelled to give an authentic setting; at the other extreme, war-gamers anxious to avoid repetitive battles can use such basic devices as simple blocks to represent hills (H. G. Wells, a pioneer war-gamer, even used books) and cottonwool to represent woods. This kind of scenery has little visual appeal of course, but it does provide a rudimentary background to the action. Most war-gamers use a compromise method between the two extremes, the hills, woods and buildings, etc., being modelled as separate flat-bottomed units, which may be placed wherever appropriate on a flat table, the latter perhaps being covered by a green cloth. This system combines the advantages of variety with a reasonably realistic appearance. Another method is to use a sand table. Terrains created in this way can look most convincing; against this, however, a good deal of space is required and sand tables can be very expensive to set up.

Model Figures

Models of figures and equipment may be built from scratch or purchased painted or unpainted. Most war-gamers use 20-mm, 25-mm or 30-mm figures. The cheapest and most widely used are of the plastic 20-mm type (00 scale); these are available at most toy shops. The more expensive metal figures are available from a number of manufacturers, some of whom are listed on the opposite page. Painted figures are usually expensive however, and war-gamers are anyway inclined to paint and remodel figures and equipment themselves. Many books are available giving details of uniforms, standards and equipment.

Dice

Dice are used in war-games to cover situations that are not completely predictable, such as the accuracy of a volley, the outcome of a mêlée, etc.

Tape Measures

Tape measures are used to determine how far a unit should be moved and to measure the range between a unit and its target when missile fire is being calculated.

Scale

Nearly all games are played within certain scales – these may vary in detail according to which of the many different sets of rules is being followed. Broadly they may be described as follows:

1 *Ground Scale.* This is the relationship between a given measurement on the table and its equivalent on a real battlefield. For example 1mm on the table may represent 1 yard in reality.
2 *Time Scale.* In reality armies would be moving and fighting continuously. This is impossible in a war-game, as all figures must be moved by hand, casualties must be calculated, etc. The game should in fact be played as a series of *cycles of play*. During each cycle of play, any or all units may move, fire, engage in close combat, and so on. Each completed cycle represents a definite amount of time within the actual battle; for example each cycle may represent one minute in reality irrespective of the time taken to play it.
3 *Figure Scale.* This term applies to the size of the figure and also to the number of men each figure is taken to represent. Even with the use of small models, the ground scale will not usually allow one figure to represent one man. Such a figure scale would not only lead to huge ranges and movements, but it would also give an unmanageably large number of figures on the table. One solution is to mount figures on bases marked to indicate the frontage and depth they represent. For example, if the figure scale is 1 figure = 20 men, each figure may represent 4 ranks of 5 men, or 2 ranks of 10 men, depending upon the army and the historical period.

Rules

Since the publication of H. G. Wells's 'Little Wars' in 1913, and R. L. Stevenson's games in the late nineteenth century, many sets of rules have been published covering many periods of warfare. War-gaming has also become far more sophisticated nowadays; for instance, one rarely sees matchsticks being fired at figures to determine hits and misses in the Wellsian manner. There has been a general movement towards greater realism, too, often at the expense of simplicity of play – though a balance must obviously be kept between playability and realism.

Most war-game rules are designed to recreate – as accurately as possible within the framework of a playable game – a type of battle which occurred at a particular point in history. All rules take into account five main aspects. These are: organization, movement, missile fire, mêlée and morale.

Organization. Figures are organized into tactical units; depending upon the period these units may be regiments, battalions, pagan war bands, Roman cohorts, etc. A tactical unit may be of any number of figures from 5 to 40, varying with the figure scale used and the period fought. Each unit is usually required to include a figure representing a unit commander and/or standard bearer. Games can be played by any number of players, each of whom controls a part of one of the armies. Each player is represented on the table by a personality figure, who is a general commander. At the beginning of the game players prepare written orders giving tactical objectives to each of the units under their control. These orders dictate the movements and actions of these units during the battle until new orders are received by the unit. Once the game begins, new orders to units are assumed to originate from the position of the personality figure. Unless the personality figure is in direct contact with the unit being reordered, a messenger figure (who carries the new orders) must be moved from the commander's position to the unit. Thus delays occur while a player decides on the appropriate response to an enemy action, and again before his counter-moves can be carried out. Verbal communication between players on the same side is also limited to those occasions when their personality figures are in contact on the table. Predetermined drum, trumpet and flag signals, and – in twentieth-century games – radio links, can be used to speed up communications.

Movement. Tables giving movement distances per period or cycle of play are based on realistic movement speeds and can be derived from the time/ground scale in use.

APPENDIX II

For example, infantry movement can be approximated to 100 yards per minute, and this would be scaled down to 100mm per game-period if the ground scale were 1 mm to the yard, and the time scale 1 minute per period. Various allowances may be made for movements uphill, through woods, across walls and marshes, etc. In some games each side makes its moves alternately, in others both sides move simultaneously. The latter system gives more realistic results, but should be used in conjunction with predetermined written orders.

Missile Fire. Scale ranges for each missile weapon can be derived directly from the ground scale used. The casualties inflicted by missiles are calculated from tables which take into account different rates of fire, accuracy, range, density of target, cover, ammunition supply, etc., etc. Once the probable number of casualties is determined, it can be varied by throwing dice – which brings a desirable element of unpredictability.

Mêlée. As in firing, tables can be used to give the relative casualties possible per period that may be inflicted by the various types of mêlée weapons and troops engaged. The tables may reflect such factors as the degree of disorganization, skill and training, armour worn, efficiency, etc., of the two forces in contact. Again the probable result is varied by dice to introduce unpredictability.

Morale. The reproduction of morale or fighting spirit is perhaps the most difficult and controversial aspect of all. Its purpose is to prevent games from becoming unrealistic slogging matches in which the loser always fights to the last man, and the victor always wins in Pyrrhic style. In reality few armies fought on after as much as 50% casualties, and most lost heart with far fewer losses. In morale tests emphasis is placed on the presence of supporting units, the number of casualties suffered, on local superiority, the security of flanks and rear, and on the results being achieved in the battle as a whole (as seen from the position of the tactical unit under consideration). Again the chance factor is indicated by a dice throw.

There are many other aspects to war-gaming. Economics, politics, religious beliefs and family connections can all be introduced as factors contributing to the way a battle or a campaign develops. However, in this short introduction it is impossible to give more than a brief glimpse of some of the aims and methods of war-gaming. The hobby, needless to say, is full of variety and provides unlimited opportunities for modelling and painting, and for historical research into uniforms, weapons, equipment, strategy and tactics, as well as the excitements and satisfactions of competitive play.

E. P. SMITH

WAR-GAME SUPPLIERS AND PERIODICALS IN THE USA

American suppliers of war-game materials include the following:

Polk's Hobby Department Store
314 Fifth Avenue, New York, NY 10001
This large retail and mail-order dealer stocks American and imported model soldiers in several scales, painted and unpainted, as well as painting and modelling accessories. Catalogue $1.49.

Scruby Military Miniatures
2044 South Linwood Avenue, Visalia, California 93277
This firm designs, produces and deals in a very wide range of unpainted metal war-gaming figures in 45-, 30-, 25- and 20-mm scales covering most nationalities and periods. Catalogue $2.00.

The Soldier Shop, Inc.
1013 Madison Avenue, New York, NY 10021
An extensively stocked shop carrying unpainted and painted flat and solid figures and a wide range of current and out-of-print military books, hard-to-find plates, recordings of military music and other hobby accessories. Catalogue $2.00.

Imrie Risley Miniatures Inc.
425a Oak Street, Copiague, Long Island, New York, NY 11726
Designers and manufacturers of 54-mm figures covering all periods.

Below is a selection of three American periodicals that aim to cover various aspects of war-gaming.

The Armchair General
P.O. Box 268, Vienna, Virginia 22180
This is a small illustrated magazine covering ancient and modern war-gaming with model soldiers. Rules, uniform information and photographs are all included in an attractive, lively format. Subscription $4.00 per annum (eight issues).

Strategy & Tactics
Simulations Publications, Inc., Room 301, 34 East 23rd Street, New York, NY 10010
A professionally published magazine, *Strategy & Tactics* offers detailed treatments of war-gaming and combat simulation. It stresses board and marker games and each issue is accompanied by a new board game covering a different period of warfare. Subscription $10.00 per annum (six issues and six board games).

The Vedette
National Capital Military Collectors, P.O. Box 30003, Bethesda, Maryland 20014
This leading American club publication includes regular features on military history, war-gaming, uniforms and model figures as well as film and publication reviews and hobby news. Large format, amply illustrated, often with a separate uniform plate. Subscription $5.00 per annum (six issues).

WAR-GAME SUPPLIERS AND PERIODICALS IN THE UK

Airfix Products Limited
Haldane Place, Garrett Lane, London SW18
Manufacturers of a very wide range of construction kits including military vehicles, buildings and scale model figures from most periods. Publishers of a monthly magazine *Airfix Magazine* (15p). This and their catalogue are available from good hobby shops and toy shops.

Hinchliffe Models
Station Street, Meltham, Huddersfield HD7 3NX
Designers and manufacturers of a very wide range of military and naval models, figures, war-game equipment and dioramas.

Hinton Hunt Figures Marketing Limited
27 Camden Passage, London N1
Manufacturers and suppliers of 20-mm, war-gaming and 54-mm collectors' figures of all periods, selling books and militaria.

Soldiers
36 Kennington Road, London SE1
One of London's largest suppliers of military miniatures, stocking over 50,000 of all makes and periods. Publishers of a monthly magazine *Miniature Warfare* (22½p).

Edward Suren
60 Lower Sloane Street, London SW1
Designer and manufacturer of 'Willie' figures, selling model soldiers, military books and specializing in dioramas.

Tradition
188 Piccadilly, London W1
Suppliers of 30-mm figures from all periods, selling military antiques, prints and books. Publishers of a monthly magazine *Tradition* (90p).

PLACES OF INTEREST TO VISIT IN THE USA

Fort Leavenworth Museum
Fort Leavenworth, Kansas 66027
Exhibits include the Von Schriltz Collection of Military Miniatures.

Fort Ticonderoga
Ticonderoga, New York 12883

Gettysburg National Military Park
Box 70, Gettysburg, Pennsylvania 17325
The battlefield of Gettysburg. Adjacent to the park is the Gettysburg National Cemetery.

Museum of History and Technology, Smithsonian Institution
14th Street and Constitution Avenue, N.W., Washington, D.C. 20560.

Patton Museum of Cavalry and Armor
Fort Knox, Kentucky 40121

Saratoga National Historical Park
Route 1, Box 113-C, Stillwater, New York 12170

West Point Museum
West Point, New York 10996

Many Civil War battlefields have become National Parks where one can see the terrain first-hand by walking over it, visit small but excellent museums and attend orientation talks or films.

PLACES OF INTEREST TO VISIT IN THE UK

Imperial War Museum
Lambeth Road, London SE1

The National Army Museum
Royal Hospital Road, Chelsea, London SW3

Tower of London

The Wallace Collection
Manchester Square, London W1

Wellington Museum
Apsley House, Piccadilly, London W1

ACKNOWLEDGMENTS

The Editors would like to express their gratitude to the following people who kindly lent their collections of model soldiers and allowed us to photograph them:
David Chandler, Peter Gilder, Charles Grant, Lt. Commander John Sandars, E. P. Smith, John Tunstill and Brigadier Peter Young.
They also gratefully acknowledge the courtesy of the following photographers, publishers, institutions, agencies and corporations for the illustrations in this volume. All maps were prepared by Ivan and Robin Dodd and drawn by Eugene Fleury, Geoffrey Watkinson and Christopher Marshall. The terrain was specially made for the book by Hinchliffe Models, Station Street, Meltham, Huddersfield HD7 3NX.

Front Flap
Philip O. Stearns

Title Page
Philip O. Stearns
The jewellery used in this photograph was specially lent to us by Harvey & Gore, 4 Burlington Gardens, London W1 and Simeon Gorlov, 19 Burlington Arcade, London W1.

Thermopylae
8 Philip O. Stearns
10 The Mansell Collection
Philip O. Stearns
11 C. M. Dixon
12–13 The Mansell Collection
13 Ivan & Robin Dodd
14–15 Philip O. Stearns
14 Philip O. Stearns
15 Philip O. Stearns
16 Philip O. Stearns

Agincourt
18 Philip O. Stearns
20 By courtesy of the Dean and Chapter of Westminster Abbey
The Wallace Collection
Radio Times Hulton Picture Library
20–21 By courtesy of the Dean and Chapter of Westminster Abbey
22–23 Philip O. Stearns
23 National Portrait Gallery, London
24–25 Philip O. Stearns
26–27 Philip O. Stearns
27 Philip O. Stearns
28 Philip O. Stearns

Edgehill
30 Philip O. Stearns
32 Radio Times Hulton Picture Library
Department of the Environment: Crown Copyright
33 The Mansell Collection
34 Photo: Michael Taylor/BPC Library
The Mansell Collection
35 Philip O. Stearns
37 Philip O. Stearns
38 Philip O. Stearns
Philip O. Stearns
40 The Mansell Collection

Blenheim
42–43 Philip O. Stearns
44 Collection Viollet, Paris
Collection Viollet, Paris
45 By kind permission of His Grace the Duke of Marlborough/BPC Library
The Mansell Collection
46–47 Philip O. Stearns
48 Philip O. Stearns
50–51 Philip O. Stearns
51 Philip O. Stearns

Lobositz
54 Philip O. Stearns
56 Ullstein
57 Ullstein
Staatsbibliothek, Berlin
58 Staatsbibliothek, Berlin
Museum of Military History, Prague
59 Philip O. Stearns
60 Philip O. Stearns
62 Philip O. Stearns
63 Philip O. Stearns
Philip O. Stearns

Saratoga
66 Philip O. Stearns
68 Anne S. K. Brown Military Collection, Brown University, Providence, Rhode Island
69 The Mansell Collection
Radio Times Hulton Picture Library
71 Philip O. Stearns
72 Copyright: The Frick Collection, New York
Anne S. K. Brown Military Collection, Brown University, Providence, Rhode Island
74–75 Philip O. Stearns
75 Philip O. Stearns

Austerlitz
78 Philip O. Stearns
80 Collection Viollet, Paris
Collection Viollet, Paris
81 Collection Viollet, Paris
Collection Viollet, Paris
82–83 Philip O. Stearns
83 Novosti
Bulloz
84 Philip O. Stearns
86 Philip O. Stearns
Philip O. Stearns
87 Philip O. Stearns

Waterloo
90 Philip O. Stearns
92–93 National Army Museum, London
92 John R. Freeman
93 By courtesy of the Wellington Museum/BPC Library
By courtesy of the Wellington Museum: Crown Copyright
95 Philip O. Stearns
96 Philip O. Stearns
98–99 Philip O. Stearns
99 Philip O. Stearns
Philip O. Stearns

Gettysburg
102 Philip O. Stearns
104 Library of Congress/BPC Library
Library of Congress/BPC Library
The Mansell Collection
104–5 Cook Collection, Valentine Museum, Richmond, Virginia/BPC Library
US Signal Corps Photo: (Brady Collection) National Archives, Washington D.C./BPC Library
105 Courtesy Chicago Historical Society/BPC Library
Library of Congress
Library of Congress/BPC Library
Library of Congress/BPC Library
106–7 Philip O. Stearns
110 Philip O. Stearns
110–11 Philip O. Stearns
111 Philip O. Stearns
113 Philip O. Stearns

El Alamein
114 Philip O. Stearns
116–17 Imperial War Museum, London
117 Imperial War Museum, London
118 Imperial War Museum, London
The Mansell Collection
119 Philip O. Stearns
122–23 Philip O. Stearns
122 Philip O. Stearns
123 Philip O. Stearns